Structure and Organization of Product Development Projects

Josef Schlattmann · Arthur Seibel

Structure and Organization of Product Development Projects

 Springer

Josef Schlattmann
Workgroup on System Technologies
and Engineering Design Methodology
Hamburg University of Technology
Hamburg, Germany

Arthur Seibel
Fraunhofer Research Institution for
Additive Manufacturing Technologies IAPT
Hamburg, Germany

ISBN 978-3-030-81048-1   ISBN 978-3-030-81046-7 (eBook)
https://doi.org/10.1007/978-3-030-81046-7

This Springer imprint is published by the registered company Springer Nature Switzerland AG
The registered company address is: Gewerbestrasse 11, 6330 Cham, Switzerland

Preface

Product development projects do not necessarily have to be complex if they are structured and organized in the right way; however, they often become complex when their manageability reaches natural limits or when coordination and project management issues are not clearly defined and comprehensible for all parties involved. In addition, leadership of employees as well as projects must be learned; otherwise, it is usually left to coincidence, and in most university education courses, it still comes up far too short.

The structure of this handbook is based on the organization and process of a product development project, such as those found in medium-sized mechanical engineering companies. The organization concerns all areas directly involved from the idea to the marketable product and includes all the necessary steps (*methodical* procedure) as well as the associated executing bodies so that the organization can be described as a *system* in its entirety and the corresponding process as *systematic*.

There are many methodical tools and guidelines for a systematic approach. The main advantages of using such an approach are, among others:

- The design process in particular can be better managed.
- The methods can be used as tools to significantly support product development.
- A goal-oriented approach creates the necessary freedom for creative action.
- The expansion of pattern formation, and especially the promotion of employee creativity, increases the probability of creative leaps.

Consequently, the importance of a systematic approach to product development cannot be overestimated.

This handbook would like to bring methods proven in industrial practice and taught in numerous lectures and seminars to a larger audience to make the innovation work of the product developer substantially more effective, particularly in small- to medium-sized mechanical engineering companies. The decisive factor here is the insight that new products are not simply generated by "designing" and "organizing" but are created by people, since behind the technical difficulties, there are, in fact, always hidden human problems that cannot be overcome by functional measures alone.

To avoid disrupting the flow of reading, we have deliberately omitted the additional formulation of female pronouns. We would therefore like to point out that the exclusive use of the masculine form should of course be understood as independent of gender.

We would like to take this opportunity to thank Prof. Dr.-Ing. Walter Jorden, the former head of the Laboratory for Engineering Design at the University of Paderborn, on whose "Paderborn Design School" this handbook is largely based. We would also like to thank the publishing house for the excellent cooperation and printing of this handbook.

Hamburg, Germany   Josef Schlattmann
Spring 2021   Arthur Seibel

Contents

Introduction to Product Development

1 General Information

Significance: The term "product development" is used to describe the entirety of activities undertaken to achieve a mature product, starting from a technical task. It comprises the two classic terms *development* and *design*, which cannot be clearly separated.

Term formation: The term product development was formed through the increasing application of a *systematic* and *methodical* approach to making the earlier more intuitive procedure for developing technical products easier to *plan* and *verify* (see Pahl et al., 2007).

Verifiability of the development: Against the background of quality assurance and changes in product liability, verifiability has gained considerable importance over the past several years. A *systematic* approach reflects, above all, a step-by-step process with correspondingly manageable steps, while *methodical* processing aims to achieve the targeted inclusion of methods (i.e., helpful tools).

Significance of the methodical procedure: The design methodology examines the course of the design process and divides it into logically successive sections so that the designer (Jorden, 1983):

- does not automatically stick with the *first best* solution (it is never the best!) but instead searches for others;
- can find solutions for *new* tasks;
- *develops* and *optimizes* existing solutions; and
- is able to *give an account* to himself and others about the design process (this is important, e.g., for documentation and product liability).

Good and experienced designers always proceed methodically—often unconsciously. Methodology helps the inexperienced learn this "art." It is a tool (not an end in itself!) that has to be adapted to each case. It strives for rationalization

© The Author(s), under exclusive license to Springer Nature Switzerland AG 2021
J. Schlattmann and A. Seibel, *Structure and Organization of Product Development Projects*, https://doi.org/10.1007/978-3-030-81046-7_1

- due to the better quality of the design results,
- as a result of greater security against errors and misdevelopments, and
- due to the division into "algorithmic" steps (computer application) and "heuristic" steps (intuition and decision of the designer).

The following statements about design methodology, however, are wrong:

- "It takes time; we can't afford it." (When used correctly, it saves time by avoiding misdirection and similar issues.)
- "It inhibits creativity." (The correct statement would be as follows: It determines only the order of the steps; their content, however, specifically allows room for creativity.)
- "It traces the design process back to logical thinking." (The correct statement would be that it involves "controlled intuition," a meaningful combination of logic and intuition.)

Performance of a product developer: While *knowledge* is based on experience and existing knowledge of properties, methods, etc. (the state of the art), the individual thinking process builds on the personal *abilities* of the product developer or the product development team. Individual abilities include spatial imagination and the ability to shape the product through analysis, abstraction, association, etc. Taken together, these abilities can also be referred to as the product developer's *skills*. These skills can be learned to a certain degree, just as creativity can be learned. From this, however, it cannot necessarily be deduced that "skills" alone constitute the product developer's performance. Rather, the *product of motivation and skills* describes the *performance of the product developer*. This can also be seen in a simple limit observation: If, for example, motivation was to approach zero, knowledge or abilities alone would hardly help. The reverse is also true: If abilities were approaching zero, then even a high level of motivation will not provide a solution. Only an extensive harmony of these parameters at the highest possible level creates a good basis for new or improved product solutions.

If the terms are broken down further, it becomes evident that the *acquisition of knowledge* depends on individual factors (e.g., intelligence). The specification of ordered and stepwise processed knowledge can considerably shorten the learning time, while skills can be improved through practice and training. The latter, in turn, requires necessary individual prerequisites. Figure 1 shows the relationships schematically.

Sources for the product developer: Basically, there are three main sources for the product developer in his work:

- *Experience:* This includes product knowledge, knowledge of production, and knowledge of risks and failures; these different types of knowledge are necessary, but there is the risk of just following well-worn tracks.
- *Inventiveness* (creativity): This is the basis of every new development or improvement.
- *Cooperation:* It is impossible for one person to know everything!

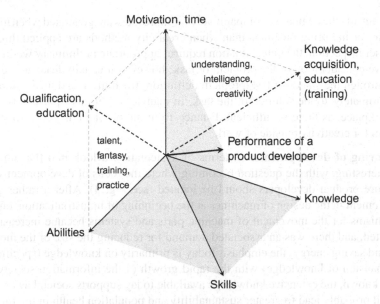

Fig. 1 Simplified model of the performance of a product developer

Finding ideas: Finding new ideas is becoming increasingly difficult. The "inventor type" is no longer sufficient. Statisticians say that only one successful product results from approximately 100 useful ideas. Approximately 80% of a company's sales growth is based on new ideas, and only 20% is based on gradual product improvements. The logical conclusion, then, is that we need

- more ideas, that is, *systematic* search, and
- better, more reliable, and faster evaluation, that is, *methodical* product development.

Location of idea generation: Statistics on the places where most ideas are generated are shown in Fig. 2. Generally, 24 of every 100 ideas are generated inside the company

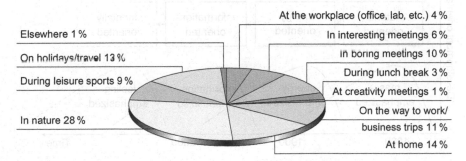

Fig. 2 Places of idea generation. Based on Berth (1993)

and 76 outside the company. Comparatively far more ideas are generated when hiking in nature or in boring meetings than when creativity methods are applied directly. On closer inspection, the latter situation requires appropriate preliminary work, with intensive consideration of the problem or task; however, it usually leads to creative leaps through the stages of task/problem definition, frustration, and incubation and thus, ultimately, to the solution of the task. In many cases, it takes place outside of the workplace, as there is sufficient distance from the problem and corresponding freedom for creativity outside of work.

Monitoring of developments: In terms of systematics, a look into the future is also interesting, with the question becoming where the focus of development work for future product developers should be located (see Fig. 3). After a rather steep improvement in the design of machines at the beginning of industrialization, energy mechanisms for the movement of machine parts and systems became increasingly integrated, and there was an associated demand for reducing the size of the moving parts and saving energy; the emphasis today is primarily on knowledge (specifically the expansion of knowledge) with the rapid growth of the information society. As is well known, the extensive knowledge available today supports societal awareness and will probably lead to greater sustainability and population health in the future.

Procedure of a product development project: A product development project is generally divided into four sections:

1) *Setting goals* (i.e., setting the *direction*).
2) *Planning* (i.e., setting the *route* to the destination).
3) *Organizing* (i.e., building and continuously adapting the *"instrument"* for realizing the plan).
4) *Executing*, which includes the following *activities*:

   - leadership (essential task!),
   - coordination, and
   - monitoring.

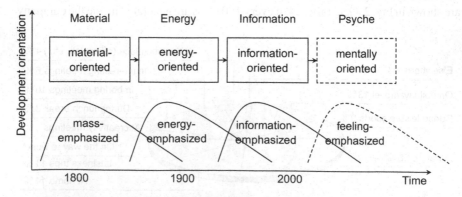

Fig. 3 Wave-shaped development of new focal points

Table 1 Leading tasks of engineers

Fields of activity	Leading tasks					
	Setting goals	Planning	Organizing	Executing		
				Leadership	Coordination	Supervision
Leading, management	× ×	× ×	×	× ×	×	×
Research, development		×	×	×	×	×
Technical consulting, sales		×	×	× ×		
Production, quality management		×	×	× ×	×	× ×
Single clerk		×				
Instructor, teacher		×	×	× ×		×

Organizational state: The product development area exists somewhere between nonorganization and overorganization:

- *Nonorganization:* Ideas can be lost, or poor or fixed ideas are pursued, etc.
- *Overorganization:* Bureaucratization, immobility, slowness ("Parkinson's principle").

The organization must be reasonably adjusted to the needs of the company in order to work as optimally as possible. One difficulty here is that there is neither a widely valid model nor a permanently optimal state. Organization is a living organism that requires constant adaptation to changing people and circumstances.

Leading tasks of engineers: These tasks can be subdivided as seen in Table 1.

2 Need for Rapid Development

Product lifecycle curve: This shows the need for the development of new products. To achieve continuous profit, a new product or significant product improvement must already have been launched (introduced) during the maturing phase; that is, the new development must start at or even before the growth phase of the product currently on the market (see Fig. 4). (The time axis is not necessarily linear, even if the orders of magnitude are approximately correct.)

Product cycle: Time from the market launch of a product to the introduction of the successor product. Table 2 provides some examples.

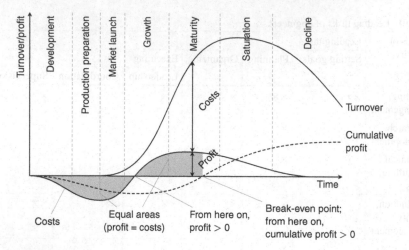

Fig. 4 Product lifecycle curve

Table 2 Product cycles of different product types (examples)

Product type	Product cycle
Fashion items	1 year
Consumer products	3 years
Large machinery	Approximately 8 to 10 years

Innovation time: Time between the invention or discovery of a working principle and its technical use; it is becoming increasingly shorter (see Fig. 5).

Possibilities for product renewal: Two strategies for a roughly continuous profit result from the life cycle (see Fig. 6):

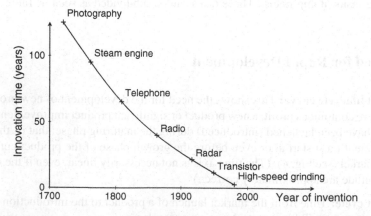

Fig. 5 Decrease in innovation times along with the year of invention. Based on Jorden (2000)

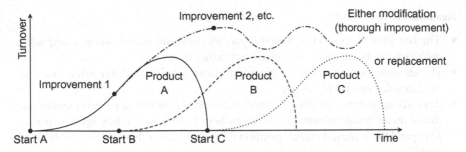

Fig. 6 Strategies for product renewal

- substantial *improvement* of an existing product or
- *replacement* of an obsolete product with a new one.

These possibilities can merge with each other. Improvement is usually possible only a few times; then, a necessary fundamental revision leads to an essentially "new" product.

Shifting of the competition: Competition used to focus on costs. Today, it is mainly about the time needed for new developments. The time required to develop a new product (i.e., from initial idea to market launch) is now longer in many areas than the average product lifetime (i.e., the time from acquisition to the product's inoperability).

Effects of time and costs: For products with a short life span (e.g., five years), the extension of the development time has a far more negative effect than an increase in development costs. Shortening the development time—even if doing so incurs increased costs—results in an earlier market launch and a steeper profit curve (see Fig. 7).

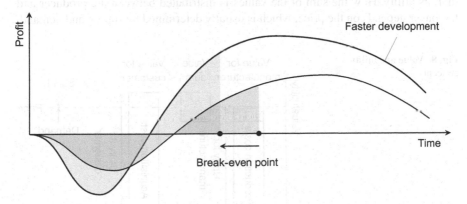

Fig. 7 Effects of a shorter development time on profit

Reasons for the effects:

- The new product enters the market *before* its competitors and thus at a time when the market is much more receptive ("act rather than react").
- Higher demand and lower competition allow a *higher market price* and thus steeper sales and profit curves.
- An early appearance enables the product to take a position as a *market leader* and thus gain an image advantage as well as better control of follow-up investments to expand the achieved market position ("innovate faster than the competition can copy").

3 Product Development Process

3.1 Planning Phase

Significance: A development and design task results either directly from a *customer order* or indirectly by *planning* carried out by company management. This product planning is carried out in a separate group independent of the development department (see Pahl et al., 2007).

Value and utility concept: Before the designer starts to design or improve a product, he must be clear about the objectives, especially about benefits and costs. The objectives can be the lowest possible manufacturing costs, the best possible function fulfillment, or a compromise that incorporates the two. The *utility* of the product for the manufacturer corresponds to the sales price; after deducting all costs, the *value* (revenue) for the manufacturer is obtained (see Fig. 8). The utility for the buyer lies in the function of the product (it can hardly be expressed directly in euros); the purchase price must still be deducted from this. No one will buy a product if its price is higher than its utility. How the sum of the values is distributed between the producer and the buyer depends on the price, which is usually determined by supply and demand.

Fig. 8 Value and utility concept

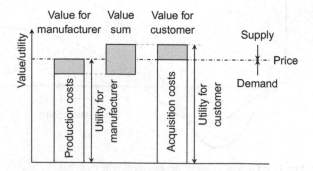

Customer and requirements specification: The customer first summarizes all requirements for the product to be developed in the so-called "customer specification"—the "what" and "why." Then, the contractor describes in the so-called "requirements specification" how he wants to technically realize these requirements—the "how."

3.2 Design Phase

Significance: According to "VDI 2221" (1993), the design phase is subdivided into "conceptual design" and "embodiment design." For the sake of clarity, however, such a division is deliberately omitted here, as it is known that the conception and design phases are closely interwoven.

Procedure: At the beginning of this phase, it is necessary to determine the abstract essence of the *task*. The abstract task formulation then results in the overall function of the product to be developed, which is then subdivided into corresponding sub*functions*. For each of these subfunctions, as many *working principles* as possible are sought, and these are then combined to form compatible principle solutions. These principle solutions are illustrated in function structures by means of corresponding block diagrams. By means of an evaluation and selection process, the large number of solutions is finally reduced to a few to be pursued. Subsequently, a modularization takes place with scale drawings of the *design elements*.

3.3 Elaboration Phase

Significance: In this phase, the geometrical and material properties of the product are defined. Here, it is often necessary to take into account a large number of design requirements, some of which are contradictory (see Fig. 9). The *overall design* finally results in production-ready manufacturing documents (drawings, parts lists, data records).

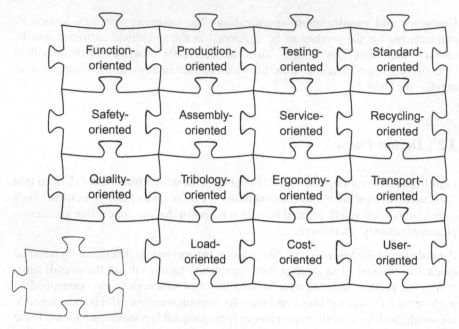

Fig. 9 Design requirements for the product developer

References

Berth, R. (1993). *Erfolg – Überlegenheitsmanagement: 12 Mind-Profit Strategien mit ausführlichem Testprogramm*. Dusseldorf: ECON.

Jorden, W. (1983). Die Diskrepanz zwischen Konstruktionspraxis und Konstruktionsmethodik. In V. Hubka & M. M. Andreasen (Eds.), *Proceedings of the International Conference on Engineering Design* (Vol. 2, pp. 487–494). Zurich: Heurista.

Jorden, W. (2000). Schrumpfen heißt Ausatmen. Analogien zum Wandel in Welt und Wirtschaft. In F. Hager & W. Schenkel (Eds.), *Schrumpfungen. Chancen für ein neues Wachstum. Ein Diskurs der Natur- und Sozialwissenschaften* (pp. 137–146). Berlin: Springer.

Pahl, G., Beitz, W., Feldhusen, J., & Grote, K.-H. (2007). *Engineering Design. A Systematic Approach* (3rd ed.). London: Springer.

VDI Guideline 2221. (1993). *Systematic approach to the development and design of technical systems and products*. Berlin: Beuth (in German).

Fundamentals of Product Planning

1 General Information

1.1 Introduction and Delimitation

Tasks of product planning: Three different tasks can be distinguished in the planning process:

- *Product search:* This constitutes the content of the planning phase and is the most important task.
- *Planning tracking:* This constitutes the control during the design and elaboration phase.
- *Product monitoring:* This begins with the time of market launch and includes:
  - monitoring the status of various areas (sales, profit, competition, trends),
  - comparing actual and planned results,
  - introducing corrective measures.

Planning periods and possibilities: According to Fig. 1, a distinction is made between:

- *short-term* planning (product/process improvement),
- *long-term* planning (new product/machines), and
- *strategic planning* (new product group/production facility).

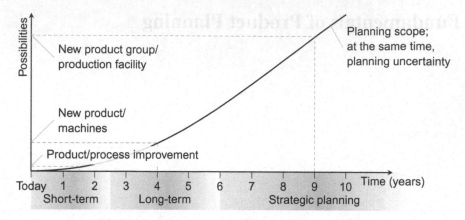

Fig. 1 Planning periods and possibilities

2 Market Behavior

Product performance and user claims: These generally do not coincide (see Fig. 2); however, the relevant components of their interaction include the following:

- *Redundant* (unused) *product services* cost money unnecessarily.
- *Uncovered claims* hold opportunities for new ideas and thus for new products or product improvements. Over time, the claims shift (the product becomes outdated).
- The task of market research is to find uncovered claims, which are influenced by the state of the art. However, they can also be specifically awakened (marketing).

Fig. 2 Product performance
and user claims

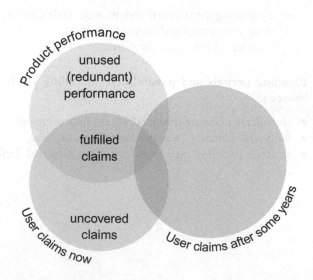

Market analysis: If the customer base is known, the company can carry out the market analysis itself. If the circle is scattered ("department store sales"), a market research institute should be hired instead. In any case, the following terms should be clearly distinguished.

Market and sales potential: See sales pyramid (Fig. 3).

Market forecasts: Some relevant terms and relationships are shown in Fig. 4.

Benchmarking: This is the direct comparison of a company with its competitors based on metrics (e.g., market share, per capita turnover, share of costs in development, and lead time in design). The benchmark ("100%") is always the best company in the reference area. The comparison can be made according to different aspects:

- *products* (industry),
- *processes* (manufacturing process, development process, etc.), or
- *resources* (potential of the company).

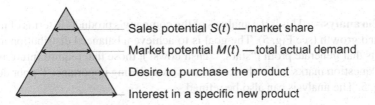

Sales potential $S(t)$ — market share
Market potential $M(t)$ — total actual demand
Desire to purchase the product
Interest in a specific new product

Fig. 3 Sales pyramid

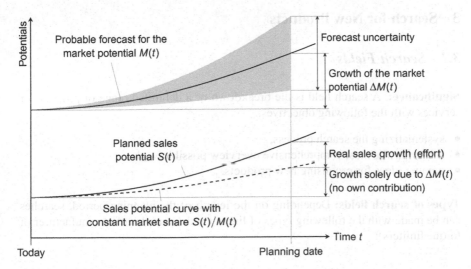

Fig. 4 Market potentials and forecasts

Fig. 5 Portfolio analysis as
four-field matrix

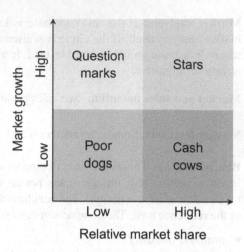

The goal is to recognize one's own weaknesses and, if possible, to match or exceed the "best in class."

Portfolio analysis: This is the analysis of the company's products in terms of market share and growth (see Fig. 5). The goal is to achieve a balanced distribution among products that generate profit ("stars," "cash cows"), those that require future investment ("question marks"), and those that are slowly losing importance ("poor dogs"), see Fig. 5. The analysis can also be refined.

3 Search for New Products

3.1 Search Fields

Significance: A search field is the breakdown of a delimited range of products or services with the following objectives:

- systematizing the search process,
- achieving the most comprehensive overview possible, and
- gaining ideas for promising new products.

Types of search fields: Depending on the length of the planning period, searches can be made with the following types of limits (with examples of a manufacturer of torque limiters):

- *short-term:* products (e.g., a new torque limiter),
- *long-term:* product fields (e.g., in the field of clutches), or
- *strategic:* functional fields (e.g., in the drive elements area).

Idea search based on market facts: The starting points are as follows:

- market requirements (e.g., customer wishes),
- the company's own deficiencies (e.g., complaints),
- the state of the art (e.g., competitions, literature, patents, trade fairs),
- future developments (e.g., technical and economic trends), and
- laws and other necessary considerations (e.g., environmental protection).

Procedure for determining search fields:

1. **Define the requirements for search fields.** These may include the following:

   - Conformity with the *company's objectives*, for example,

     - ensuring the company's continued existence,
     - expanding the company's flexibility (customers, product groups),
     - reinforcing the company's corporate image (e.g., luxury or department store items).

   - *Future viability*, as seen in Table 1.
   - Matching the *company's potential* (see Table 2).

2. **List possible search fields.** These are listed using a hierarchical structuring of the area under consideration (e.g., "measuring devices"). Thus, one should proceed:

   - more pragmatically than strictly logically (there is no "strictly logical" outline, because several aspects of order intersect with each other), and
   - from the general to the specific.

3. **Select a promising search field.** This is done according to:

   - the requirements developed in Point 1 or
   - the nature of the planning period (short-term/long-term/strategic; possible expansion of the current product range).

Table 1 Influencing factors on future viability (selection)

Market development	General trends	Technological trends	Substitution
Customers Competitors Procurement market	Economic environment Ecological environment Population development Consumer behavior	New processes New working principles New materials	Of the task Of the function fulfillment

Table 2 Potential opportunities in the company

Type	Area			
	Development potential	Production potential	Sales potential	Procurement potential
Information potential	Experience Property rights Technical literature	Experience Organizational structure	Experience Sales organization Customers	Experience Organization Suppliers
Equipment potential	Test facilities Measuring instruments Prototype production	Buildings Machines Infrastructure	Branches Equipment Transportation	Equipment Transportation
Staff potential	Research staff Design engineers Toolmakers	Skilled workers Support staff	Indoor services Field services	Indoor services Field services

3.2 Diversification

Significance: Diversification means the expansion of the product range based on existing products, facilities, and experience. The following search directions can be used systematically in diversification efforts:

- *Process:* Which new products can we manufacture with our existing production facilities and experience?
- *Customers:* Which products can we sell to the existing customer base or with the existing sales organization?
- *Material:* Which products can we manufacture from existing materials and semifinished products?
- *Principle:* What products with the same or similar working principle can we manufacture?
- *Function:* Which products with the same or similar overall function can we manufacture?
- *Production depth:* Which products can we manufacture that are related to existing products in terms of vertical integration (e.g., vendor parts, extras, assembly with other products, machines for manufacturing or processing of our products)?

The last item is also called "vertical" diversification; the others fall under "horizontal" diversification.

Practical example: A chemist originally built and sold flue gas analysis equipment for chimneysweepers. This equipment eventually came to include the following:

- chimneysweeper services (customer area),
- various brushes (manufacturing process, function, material for one of the chimneysweep products), and
- brush manufacturing machines (vertical diversification; current main product).

4 Improvement of Existing Products

4.1 Value Analysis

Significance: A special design methodology that is well established and proven in practice is value analysis ("VDI 2800 Part 1" 2010). Here, thinking about functions and costs per function plays an important role. A distinction is made between the following:

- functions according to their meaning for the *task*,
- functions according to their meaning for the *user*.

Functions according to their meaning for the task (see Table 3 for examples):

- *Main function:* This function is directly necessary to fulfill the task. It can be equal to the overall function but does not have to be.
- *Secondary function:* This function enables or supports the main function. It can be equal to a subfunction but does not have to be.
- *Undesirable function:* This function exists but is neither useful nor desirable. It must be detected and, if possible, removed.

Functions according to their meaning for the user (see Table 4 for examples):

- *Usage function:* This function is required for use and can be quantified objectively.

Table 3 Functions according to their meaning for the task using the example of a light bulb

Main function	Emitting light
Secondary function	Holding bulb in socket
Undesirable function	Emitting heat

Table 4 Functions according to their meaning for the user using the example of a light bulb

Usage function	Emitting light
Prestige function	Helical drop shape
Exchange function	"Made in Germany," brands
Function-related property	Power 40 W

- *Prestige function:* This function satisfies non-use-related needs and can be stated only subjectively.
- *Exchange function:* This function motivates the user (buyer) to prefer one product over an otherwise equivalent competitive product.
- *Function-related property:* All functions mentioned are qualitative in nature. Their quantification, that is, their corresponding numerical value, is also referred to as "function-related property" in value analysis.

Success of the value analysis: It is based primarily on the practice-oriented inclusion of *organizational* rules (see "VDI 2800 Part 2" 2010) and supported by *methodical* tools (see the following two sections).

4.2 ABC Analysis

Significance: ABC analysis is a formal tool for listing and sorting the manufacturing costs of a product's individual parts and, if necessary, modules as well as the associated assembly procedures. Thus, the main cost areas and starting points for cost savings can be identified using ABC analysis.

Procedure: This is explained using the example of the protective hood seen in Fig. 6. The steps correspond to the columns in Fig. 7.

1)   Enter the *position numbers* consecutively.
2)   List the *function carriers*; these can be

   – individual parts,
   – preassembled modules (for larger units), or
   – assembly operations (if not included in modules or, if applicable, individual parts).

Fig. 6 Protective hood

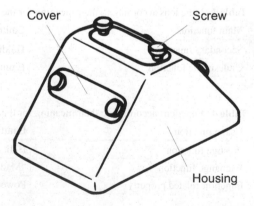

Cover

Screw

Housing

Function carriers listed						Sorted by rank					
Position number	Name	Quantity		Single costs h (€)	Costs in product				Quantity	ABC group	Position number
		n	%		n · h (€)	%	%	Cumulated			
								%	%		
①			②	③	④	⑤	⑥	⑦	⑧	⑨	⑩	⑪	⑫
I	Cover	3	30	3	9	18	16	16	10	A	2
2	Housing	1	10	38	38	76	18	94	40	B	1
3	Screw + assembly	6	60	0,5	3	6	6	100	100	C	3
	Sum	N = 10	100		H = 50	100	100				

Fig. 7 Form for the ABC analysis

3) Specify the *quantity per function carrier n* in the product; calculate the *total quantity N*.

4) Enter the *percentage quantity* $(n/H) \cdot 100\%$ (sum = 100%).

5) Calculate the *manufacturing costs per function carrier h*.

6) Enter the *function carrier costs* $n \cdot h$; calculate the *total manufacturing costs H*.

7) Enter the *percentage costs* $(n \cdot h/H) \cdot 100\%$ (sum = 100%).

8) *Rank* the percentage costs in Column 7 by entering the highest value first. Enter the corresponding *position number* (from Column 1) in Column 12.

9) Enter the *cumulative percentage costs* by summing up the values from Column 8.

10) Enter the *cumulative percentage quantity* by taking the values from Column 4 according to the item number (Columns 12 and 1) and summing them up.

11) *ABC allocation:* If the costs from Column 9 are plotted over the quantities from Column 10, this usually results in a distribution similar to that seen in Fig. 8. Only a few expensive parts and/or procedures make 75 to 80% of the proportional manufacturing costs (A parts); here, a cost reduction would be profitable. For B parts (approximately 95% of the costs), a cost reduction would be questionably profitable; above that (C parts), such a reduction would not be recommended. These limits are not strictly defined.

4.3 Function–Cost Matrix

Significance: A product should fulfill certain functions. Its manufacturing costs are determined by certain function carriers (see Fig. 7, Column 2). The question of which function incurs which costs may be essential for the evaluation and improvement of the product. Many function carriers are involved in several functions. Therefore, the function carrier costs (see Fig. 7, Column 6) may have to be split between several functions. This is done by estimating the shares (in percentage).

Fig. 8 Principle of the ABC assignment

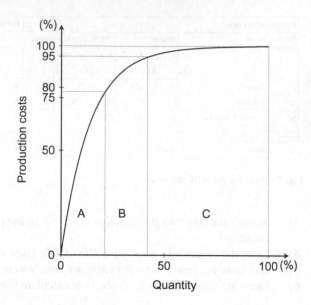

Procedure: The procedure can also be explained using the example of the protective hood from Fig. 6. For this purpose, the form in Fig. 9 is used. Columns 12, 2, 3, and 6 (the numbers are taken from Fig. 7) show the function carriers sorted by rank. Often, C parts can be omitted for simplification purposes. Steps 13 to 16 follow the numbers from Sect. 4.2; both forms can be combined into one form if necessary.

12) Enter the *function numbers* into the form (see Fig. 7, Column 1).
13) List the *functions*.
14) Assign the *type of function* and number it if necessary (this step may be omitted), for example:
   - M = main function,
   - S = secondary function,
   - U = unwanted function.

Function carriers				Functions									
Position number	Name	Quantity n	Costs in product $n \cdot h$ (€)	Name (13)	Protect environment		Enable opening/closing		Look good		Provide stability		
(12)	(2)	(3)	(6)	Type/no. (14) Share (15)	M1 g (%)	gnh (€)	M2		S1		S2		
2	Housing	1	38		60	22.80	25	9.50	5	1.90	10	3.80	
1	Cover	3	9		35	3.15	60	5.40	5	0.45	-	-	
3	Handle screw	6	3		20	0.60	60	1.80	20	0.60	-	-	
				Function costs (16)		26.55		16.70		2.95		3.80	

Fig. 9 Form for the function–cost matrix

15) Estimate the *function carrier shares* of the functions (row by row). The following questions can help in doing this:

   - Which function carrier has a share in the fulfillment of which functions? (If none or low, cross out the matrix field).
   - What percentage of the function carrier costs $n \cdot h$ is caused by the function in question? Alternatively, by what percentage would the function carrier become less expensive if the function in question were eliminated?
   Estimate the percentage of g but do not exaggerate the accuracy of the estimate. In case of doubt, first assume an even distribution; leave out small shares.
   The g values in each line must equal 100%. The value of $g \cdot n \cdot h$ in each row is equal to the costs of $n \cdot h$ for the respective function carrier.

16) Calculate the *function costs* h_f for each function by finding the value of $g \cdot n \cdot h$ in each column.

Meaningfulness: The function–cost matrix helps answer the following questions:

- How much does it cost to fulfill each individual function?
- Do the costs correspond to the *importance* of the function? (If necessary, overly expensive functions should be economically improved or omitted by asking, "Is the function even necessary?")
- Which function carrier fulfills multiple functions? (This is often inexpensive.)
- Which function requires several function carriers? (This can be functionally advantageous, but expensive.)

The function–cost matrix represents only an estimation. Therefore, the level of detail should not be exaggerated.

References

VDI 2800 Part 1. (2010). *Value analysis.* Dusseldorf: Beuth (in German).
VDI 2800 Part 2. (2010). *Value analysis workplan according to DIN EN 12973 – Form set.* Dusseldorf: Beuth (in German).

Organization of a Product Development Area

1 Organizational Structure

1.1 Structure of a Company

Overview: Almost every company has some form of the organizational features shown in Fig. 1.

Structuring principles: The following applies to the relationship between the structuring characteristics "functions" and "products":

- The seven functional areas are always present in one form or another (if necessary, they are summarized or structured differently).
- A subdivision according to products is also usually present (except in the case of specialized companies with only one product).
- Problems can arise if both principles are confused (often in long-lived organizations; this results in nested organization charts).

Standard cases: Borderline cases for the dominance of a structuring principle are the following:

(a) If all functions are combined in house, the functional classification dominates, and the product classification has only an internal effect; for example, this may occur when the products are subdivided into groups.
(b) If there are several plants, product classification dominates (division of products into different plants), and the functions are then found in the organization of the individual plants.

© The Author(s), under exclusive license to Springer Nature Switzerland AG 2021
J. Schlattmann and A. Seibel, *Structure and Organization of Product Development
Projects*, https://doi.org/10.1007/978-3-030-81046-7_3

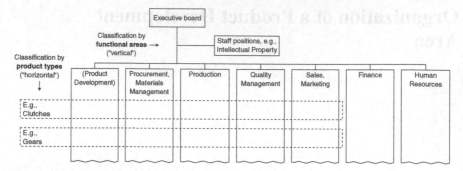

Fig. 1 Model structure of a company

Guiding rules for the organizational structure:

1 Unambiguity

The assignment must be unambiguous; that is, each unit is directly subordinate to only one superior unit.

2 Logic

The organizational structure must be logically clear.

3 Pragmatics

The organizational structure must take into account the existing circumstances (persons, premises).

Guiding Rules 2 and 3 are often contradictory; in such cases, a reasonable compromise must be sought.

1.2   Structure of a Product Development Area

Overview: Basically, the same principles and basic rules as those in the company as a whole can be found in the product development area (see Fig. 2).

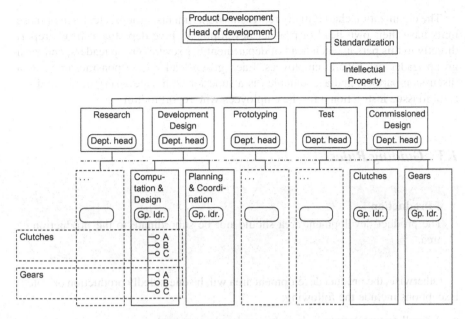

Fig. 2 Model structure of a product development area

Departments: The departments in Fig. 2 include all the essential functions of a product development area:

- *Research* is usually found only in larger companies.
- *Development design* deals with new or significantly improved products, depending on market requirements or similar forces (e.g., series products sold from stock), but generally not for individual customers.
- *Prototyping* produces prototypes. This area is not always directly incorporated into the product development area but should be part of it. If prototypes are produced in the production or tool making area, there are usually difficulties in meeting deadlines (orders from production are under time pressure). Fast access, not the utilization of machines, is decisive for effective development.
- *Test* is almost always present.
- *Commissioned design* mainly processes offers and orders for customers. (Both are always under time pressure.)
- *Quality assurance* can also be included if necessary.

Organization chart: This chart shows the assignment of personnel responsibility. Since each position may have only one superior, a tree-like structure is created. Individual employees are generally not listed; at most, their number in the organizational unit is given. (The subdivision of the computation and design group indicated in Fig. 2 also does not belong in an organization chart.)

The organization chart initially represents departments, groups, etc. Each of these units have their own head of product development. Five department heads report directly to this person; each head of department has several group leaders, and each group leader has several employees. Each group leader is responsible (e.g., as a discussion partner) and accountable (as a superior with supervisory duties and the right to issue instructions) for the employees who report to him.

1.3 Guiding Rules

4 Production

The product development area should not be subordinate to the production area.

Otherwise, the product development area will be one-sidedly production oriented. Exceptions include the following:

- A small company with little in-house development.
- Equipment design is part of production.

However, it is fitting if both areas are adjacent to each other under a "technical director" (or similar).

5 Prototyping

The prototyping department should not be subordinate to the production area.

The latter is usually under deadline pressure; development orders can be postponed if doubts arise.

6 Design

The development design and commissioned design departments should be separated if possible.

The reason for this is similar to Guiding Rule 5, as the commissioned design department faces deadline pressure.

7 Number of employees

A supervisor should not be responsible for more than five line members and, if necessary, two other staff members, that is, a total of no more than seven employees.

Otherwise, there will not be enough time for the supervision of the individual employees.

8 New organizational level

If the number of employees grows beyond seven, an additional organizational level should be created.

Human problems can arise if a former colleague becomes a supervisor or an external candidate is hired; in the latter case, existing employees may see themselves as deprived of an opportunity for advancement.

2 Responsibility and Communication

2.1 Line and Matrix Organization

Line organization: In this form of organization, the instructions come from the respective line manager according to the organization chart (see Fig. 3).

For example, the project "Clutch C11" passes through all departments. A member of the design department was appointed as the project leader. His instructions have

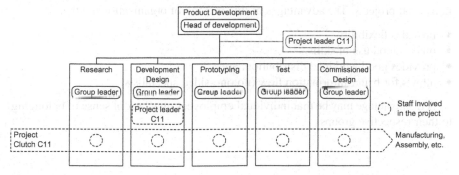

Fig. 3 Model structure of a line (dashed line) or staff line organization (project leader indicated with dotted line)

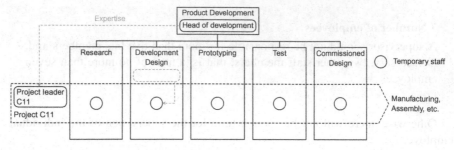

Fig. 4 Model structure of a matrix organization

to go through the "official channels" (see below). This form of organization is unambiguous but cumbersome and prone to failure.

Staff line organization: The project leader is a staff member in the product development area. This gives him more freedom and a closer connection to the head of development but does not give him any direct authority to instruct the departments involved.

Matrix organization (temporary team formation): The project leader is assigned employees directly from the individual departments concerned, precisely defined as *persons* involved in a specific *project* for a defined period of *time*. The project leader is authorized to give instructions to the employees in areas of factual or logistical concern regarding the project. For all other questions (e.g., salary), the line supervisor remains primarily responsible, but he has no authority over the project. In the event of a conflict of interests, the next highest line supervisor is responsible. Within intensive teams, the team leader tends to be given more authority, that is, he is involved in decisions regarding, for example, performance appraisals. The matrix organization is illustrated in Fig. 4.

Pool formation: In this project staffing format, employees are not permanently assigned to individual groups but form a pool from which they are assigned to individual projects. The advantages of this method of organization are that it

- provides flexibility,
- makes good use of skills,
- provides good workload distribution, and
- allows for better information flow (no one-sided specialists).

A disadvantage may be that individual employees may lack of sense of belonging to their respective groups.

2.2 Communication Channels

Significance: Without communication (conveying information), no organization can function. Consequently, communication is an important leadership task. Information can flow in different ways within an organization. Figure 5 shows an example:

- *Procedure 1:* Designer A wants to inform assembler L that a certain bearing has different seals due to urgent customer requirements and therefore has to be mounted differently (more carefully).
- *Procedure 2:* Designer A wants to inform all special bearing assemblers that, due to a design change, all bearings of a certain type series must be mounted in a slightly different way.

The possibilities for information transfer methods include the following:

- "To": information is given directly to the addressee.
- "FYI" (for your information): information is simultaneously given to others; this is parallel to "to."
- "Via": information is given to one person after another via intermediate line points.

Chain of command: The message runs through all line instances (superior units); cross information is possible only between those instances that report directly to the same superior units. This means that all affected units are informed. The disadvantages of this procedure are as follows:

- It can take some time until the message arrives;

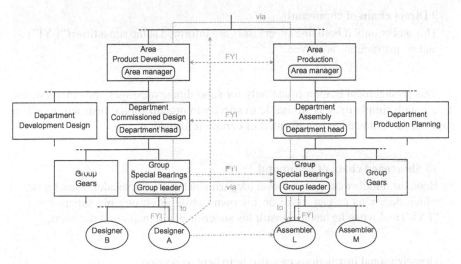

Fig. 5 Example of communication channels within an organization: normal chain of command (dots), direct chain of command (dashed), shortened chain of command (dashed and dotted)

- there is a risk of the message being stranded/overlooked; and
- there may be an unnecessary amount of information shared.

Direct chain of command: In this case, the message is sent directly to the addressee, thus ensuring the fastest possible delivery process. The disadvantages of this formation are as follows:

- There is a risk of uncontrolled measures (the supervisor is not informed or involved);
- there may be a lack of information from higher authorities with a greater overview of the situation; and
- a body not authorized to issue instructions may directly initiate communication.

Shortened chain of command: In this case, the information travels via the direct supervisors of the sender and the addressee, with information simultaneously being sent to the next highest superiors ("FYI"). The advantage of this arrangement is a relatively fast flow of information with the direct involvement of the supervisors and simultaneous dispersal of information to all involved parties. A disadvantage here is the risk of measures being implemented before a higher authority (with greater experience) can raise an objection. Overall, this arrangement is the least disadvantageous, provided that responsibility has been sensibly delegated (see the following guiding rules).

2.3 Guiding Rules

9 Direct chain of command
This works only if both line supervisors are informed at the same time ("FYI") and no instructions are given

This arrangement is appropriate only for subordinate messages and emergencies. To aid with simplicity, it is advisable to specify in organizational instructions when superiors do not need to be informed of certain information.

10 Shortened chain of command
Here, the middle management level (department head/group leader) must know which decisions he can make on his own (while informing his superior via "FYI") and when he has to consult his superior before making a decision.

Organizational instructions can also help here as needed.

11 Information "down"
When providing information to subordinate authorities down the line, especially in the case of instructions, never pass over an intermediate level.

Information should be sent using the "via" or "FYI" methods. In any case, instructions should be discussed with the superior of the addressee. Otherwise, there may be trouble with the latter if a higher superior intervenes directly in his area.

12 Line functions
The organization chart shows line and staff functions in their assigned tree structure but not in their time-related team formation (matrix organization).

The project leaders, for example, always have line supervisors.

13 Organizational units
Basically, the organization chart contains only organizational units (departments, groups, etc.), but not positions or position holders.

The position holder who manages the organizational unit in question can also be named.

3 Simultaneous Engineering and Lean Production

Former project sequence: In the project sequence outlined above, the activities of the individual areas or departments always follow one another ("sequentially"). In practice, this often led to "mental walls" between the departments, which can inhibit the flow of information and cause the development process to expand over time (see Ehrlenspiel & Meerkamm, 2013).

Disadvantages of the former project sequence: The product developer/designer defines the product along with all its details. Although he will try to take into account the concerns of the subsequent areas (e.g., consider production, costs, market, and environment concerns in the design process), he is *not* able to sufficiently oversee everything and optimize it, especially with regard to rapid technical developments (robot technology, new materials, optimized simulation software or options, etc.). He would have to constantly consult experts or expert systems. However, such consultations would often have to be omitted because of time constraints; the result would be half-hearted and insufficiently thought-out solutions. As time progressed, these

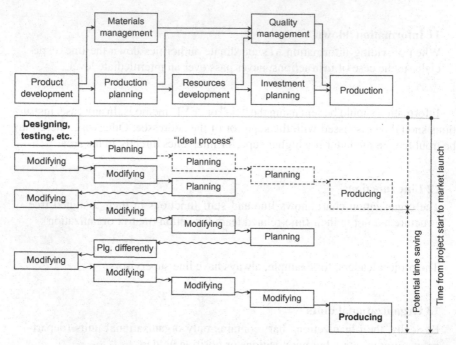

Fig. 6 Example of a change chain in a "sequential" project sequence

solutions would lead to identified deficiencies and increased costs or time-consuming and expensive changes (see Fig. 6).

Each area generally has its own customs (its own "language"), descriptive models, etc. The interface is sometimes not sufficiently illuminated, which can lead to susceptibility to errors or wasted time.

The more the work process is divided among different specialists, the more the individual loses sight of the overall objective and the less he feels responsible for the activities of others. The sum of the optimizations in the individual departments is by no means equal to the optimization of the overall process.

Today's project sequence: Instead of a sequential process, it is necessary to actively and simultaneously involve individual areas and departments during product development. Ideally, the employees of all involved departments would work in the same room all the time.

14 Simultaneous Engineering

This comprises a *joint*, *simultaneous*, and *trustworthy* performance of all engineering activities for the development of a *product* and the associated *production facilities*.

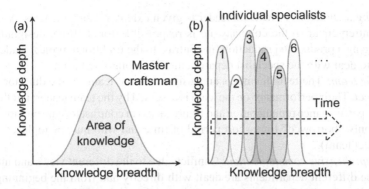

Fig. 7 Illustration of knowledge areas: **a** handcraft production, **b** mass production

Historical classification: "Simultaneous engineering" is closely related to the terms "lean production" and "lean management" (see Pfeiffer & Weiß, 1994). These terms originated in different places (simultaneous engineering originated in the USA and lean production in Japan). Lean production refers to the entire field of development and production. It emerged in three phases:

- *Handcraft production* (about until 1900)*:* Development and production occur in one location. The master craftsman determines the product and how it is manufactured, modified, and improved. The work is optimally targeted and adaptable. A limit is reached when a product becomes too complex or the number of pieces too great. Then, development and production require such a large "area of knowledge" that one person (or even a few) is not able to complete them alone (see Fig. 7a).
- *Mass production* (starting around 1900)*:* This type of production is based on the Taylor principle of the division of labor ("Taylorism"). A series of specialists (experts) with a narrow but in-dept knowledge cover of a large area of knowledge together but with a time lag (see Fig. 7b). Problems in this type of production occur at "interfaces" and "departmental walls." A limit is reached when time and flexibility become decisive factors.
- *Lean production* (from about 1990 on)*:* This type of production is based on the craft principle (an overview of the whole and an orientation of all efforts toward the visible common goal). Thus, simultaneous coverage of the ever-growing necessary knowledge area is achieved by a team of experts with integrated computer technology.

Characteristics of simultaneous engineering and lean production: Neither can be clearly distinguished from the other. They cover the entirety of all development and production areas involved, and the following techniques may be used in either system:

- *Strong team leader:* The team leader is given extensive competencies. As a kind of entrepreneur in the company, he is responsible for drafting, designing, and bringing a product to production (in contrast to the traditional project leader, who has to deal with the specialist departments on an ongoing basis).
- *Close team:* The team members are assigned to the team for the duration of the project. Their performance on the team is assessed by the team leader; it influences their performance remuneration and future tasks (in contrast to team members who are only "borrowed" for a short period of time and who may be replaced on the project team).
- *Comprehensive communication:* Conflicts due to the different views and interests of the different departments are dealt with directly and from the beginning. The decisions made together must then be actively supported by all. This requires the highest effort at the beginning of the project (in contrast to the earlier, rather sequential process in which conflicts often appear only toward the end of the project).
- *Simultaneous development:* The individual development activities overlap and do not follow each other in time (see Fig. 8). For example, the design of press tools for car body parts begins shortly after the start of the car body design. This saves considerable time but requires joint work with visual contact, experience, and foresight.
- *Involvement of the sales department:* The sales department brings the results of market research, service and customer surveys, and competition and fault analyses directly into the development work and monitors their implementation.
- *Cooperation with suppliers and manufacturers of production equipment:* These outside actors generally participate in the project work for a fee. The decisive factor is that trust-based cooperation is established (in contrast to the traditional approach, where external companies are often pressured and played off against each other).
- *Group production:* In lean production, comprehensive activities are carried out in individual groups (e.g., the assembly of an entire product, as opposed to splitting

Fig. 8 Typical project sequence according to simultaneous engineering

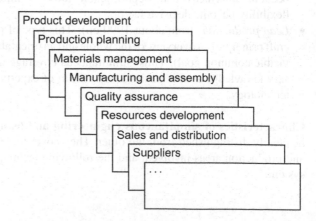

Product development
Production planning
Materials management
Manufacturing and assembly
Quality assurance
Resources development
Sales and distribution
Suppliers
. . .

up on the assembly line so that an individual only tightens three screws before passing the part on). The progress of work and the results are visible, give a sense of achievement, and lead to individual workers' personal identification with the product.

References

Ehrlenspiel, K., & Meerkamm, H. (2013). *Integrierte Produktentwicklung* (5th ed.). Munich: Hanser.
Pfeiffer, W., & Weiß, E. (1994). *Lean Management*. Berlin: Schmidt.

Structure and Process of a Product Development Project

1 Model Flow Chart

Basic scheme: A product development project can be described schematically in a simplified form using a "3 × 3" matrix (see Jorden & Weiberg, 1977). This matrix consists of three levels ("processing," "controlling," and "deciding") and three successive phases ("planning," "designing," and "elaborating"). While the processing level can extend over the entire operation, the control function is usually the responsibility of a project leader who has been specifically designated to lead the task. The project leader coordinates and monitors the individual activities within the phases. The final decision at the end of a phase is made by corporate management or a committee appointed by corporate management. The basic scheme of a product development project is shown in Fig. 1.

Model flow chart: The basic scheme from Fig. 1 is transferred to an exemplary flow chart, as shown in Fig. 2, which may correspond to the project flow in a medium-sized mechanical engineering company with a series of products. This plan is the basis for a consideration of typical individual activities in the development process. The responsible participants in the flow chart are as follows:

- E—(head of) product development,
- F—(head of) production,
- G—corporate management,
- L—project leader,
- M—(head of) materials management,
- P—(head of) product planning,
- Q—(head of) quality management,
- V—(head of) sales.

Differences from "VDI 2221 (1993)": In contrast to the model flow chart in Fig. 2, "VDI 2221" (1993) contains four phases; specifically, the design phase is divided into "conceptual design" and "embodiment design." Depending on the task, however,

© The Author(s), under exclusive license to Springer Nature Switzerland AG 2021   37
J. Schlattmann and A. Seibel, *Structure and Organization of Product Development Projects*, https://doi.org/10.1007/978-3-030-81046-7_4

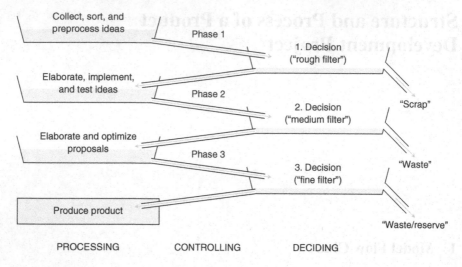

Fig. 1 Basic scheme of a product development project

the emphasis varies, and the boundaries are fluid. Therefore, three phases are recommended as a basis; if necessary, a further subdivision can be made in individual cases. The three levels are organizational specifications and do not appear in "VDI 2221" (1993).

Transfer to practice: The model flow chart must be tailored to the needs of the respective organization. For this purpose, the content of the phases and especially of the individual numerical steps can be defined differently. In any case, however, the basic "3 × 3" scheme should be retained as a minimum size, even in smaller companies. Further intermediate decisions can be inserted as required.

Ongoing adaptation: Since a given organization is constantly in a state of flux, the form of project monitoring implemented must also be constantly reviewed and adapted. This requires a determined effort on the part of both the development management team and the employees at the control level. Without vigilance, methodical product development tends to "fall asleep" and thus become disorganized. Apparent savings in project management usually result in multiple costs (=lost profit) at other points in the project process. Even smaller companies should have at least one employee who is involved in project management full time.

2 Activities in the Model Flow Chart

General: In the following, the phases listed in the model flow chart in Fig. 2 and examples of individual activities will be examined in more detail using the line

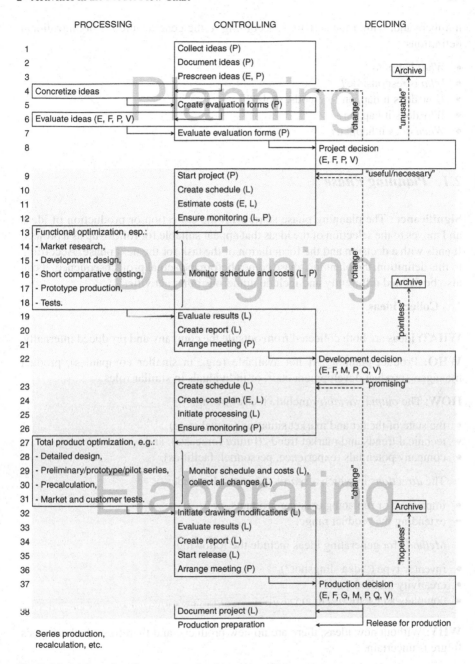

Fig. 2 Model flow chart for a product development project

numbers and, within the activities, according to the general rule for organizational definitions:

- *What* happens?
- *Who* is responsible?
- *How* does it happen (methods, tools)?
- *Why* does it happen?
- *When* does it happen?

2.1 Planning Phase

Significance: The planning phase starts with the collection or production of ideas and moves to the selection of the ideas that appear suitable for follow-up as a project. It ends with a decision and the formulation of the task for the next phase. According to this definition, the planning phase comes before the actual project sequence. It can also be defined differently and include numerous other activities.

1   Collect ideas

WHAT: Ideas are both collected from outside the company and produced internally.

WHO: Product planning; if not available (e.g., in smaller companies), product managers, product development staff, or individuals in similar roles.

HOW: The *output variables* include the following:

- the state of the art and market situation (actual state),
- technical trends and market trends (future forecasts), and
- company potentials (experience, personnel, facilities).

    The *directions* in which a *search* can extend include:

- improving or replacing existing products, or
- extending the product range.

    Methods for generating ideas include the following:

- inventor type ("idea slingshot"),
- creativity methods, and
- company suggestion schemes.

WHY: Without new ideas, there are no new products, and therefore, the business's future is uncertain.

WHEN: Basically, permanent; creativity groups are formed about every four to six weeks.

2 Document ideas

WHAT: There must be a clear, organized procedure so that ideas do not simply disappear into a "drawer" and information and suggestions can be recalled at any time.

HOW: Idea cards or sheets may be useful (see Fig. 3 for a template):

- index cards (DIN A4 or DIN A5),
- sketch sheets (DIN A4 paper or electronic).
- The cards or sheets should be stored in a database or in a similar manner (storage is important).

Examples of *order schemes* include the following:

- P—product ideas:

  - P-A for product group A (possibly subdivided),
  - P-B for product group B.

P	Title:		Product group:		
V			Date:	No.	
Drawing:			Creator:		
O					
Brief description:					
O					
Remark:		E	F	P	V
Business potential					
Market potential					
Production potential					
Development potential					
Totals					

Fig. 3 Template for documenting ideas

- V—process ideas:

  - V-A for process group A (as with the product ideas).
  - The groups can be sorted by date and location; the latest idea (sheet of paper) can be placed on top or in front, etc.

The *reasons* for using such techniques include the following:

- Product or process ideas can be pursued by different areas (development or production).
- Individual groups of ideas are easy to keep track of, and suggestions are possible, for example, when looking through the ideas.
- "Slow-moving ideas" are at the bottom of the list or can be found in an assigned category.
- An understandable (characteristic) key is easier for humans to handle (e.g., P-A instead of 1.2).

3 Prescreen ideas

WHAT: Looking through the newly arrived ideas.

WHO: Head of product planning (or similar actor), together with head of product development (it is important not to delegate this task).

HOW: Joint discussion.

WHY: The *tasks for a preliminary viewing* include the following:

- From the beginning, untenable ideas are dropped (to the relief of the evaluation group, see Point 5). However, beware! Sound justification is necessary (e.g., a product development project with a report or similar deliverable); otherwise, there is a risk of making the wrong decision or abuse of the system. In case of doubt, always forward the idea.
- Unclear or overly global ideas should be clarified or narrowed in consultation with the originator.
- Future ideas may have to be technically specified (sketched, calculated, estimated, computer simulated) before a meaningful evaluation can be made.
- Ideas such as suggestions for improvement must always be submitted to a responsible body (e.g., a person of trust).
- Ideas that could involve innovation in terms of novelty and content should always be submitted to a competent authority. In case of doubt, always decide in favor of pursuing the idea.

WHEN: This should occur every one to two months (this should be specified), depending on the ideas received.

4 Concretize ideas

WHAT: The ideas may touch on very diverse content, from the clarification of simple factual questions to a comprehensive product strategy. Thus, the "what" can include, among other things,

- market research (market size, space, trends, buyer behavior, legislation, etc.),
- cost planning (market price, development costs, investments),
- time planning (time required for market launch),
- marketing planning (sales, advertising, etc.).

WHO: For more extensive planning activities, a separate product planning department, which often belongs to the sales area, is usually responsible.

5 Create evaluation forms

WHO: Representative of the product planning department.

HOW: *Recommendations* include the following:

- Duplicate evaluation sheets or email them to the evaluation group (see Point 6); there should be one sheet per idea (if necessary, more information can be included on the sheet, see Point 2).
- Set a deadline for the completion or return of the sheets.
- Send the sheets to the members of the evaluation group (confidentially).

WHEN: This should occur every three to six months, depending on the number of ideas (this should be specified). This cycle also applies to Points 6 to 8 below.

6 Evaluate ideas

WHO: The *evaluation group* includes:

- head of product development (E),
- head of production (F),
- head of product planning (P), and
- head of sales (V).

Alternatively, competent decision-making staff may be appointed.

HOW: For the *evaluation scheme*, see Fig. 3,

- An example of a points scale would be: 4–ideal, 3–good, 2–average, 1–bad, and 0–unrealistic.
- Each evaluator enters the points according to his evaluation in the respective column of the documentation template and then sends the evaluation back.
- A "0" can mean—depending on the evaluation scheme—that the respective idea is regarded as not executable.

7   Evaluate evaluation forms

WHO: Representative of the product planning department.

HOW: *Recommendations* for this are as follows:

- Transfer point totals and values of the individual evaluators for each idea to a common sheet.
- Take out ideas with ratings of 0 (they block the idea process and do not need to be discussed).
- Select the ten highest rated ideas from the products and processes groups.
- Send selected ideas to the project decision group.
- Schedule a project decision meeting.

WHEN: This should occur immediately after each evaluation action (see Points 5 and 6).

8   Make project decision

WHAT: Determine which idea should be pursued. (The rest is documented in the archive or electronically; see Point 38). Formulate the *task* for the *design phase* as precisely as possible, including determining the following:

- keywords for the function, etc.;
- necessary investigations (e.g., market data), if not already done (see Point 4);
- time frame for the design phase (see Point 10); and
- a cost framework for the design phase as well as product manufacturing costs (see Point 11).

WHO: Project decision group, for example, consisting of:

- head of product development (E),
- head of production (F) (essential for production ideas),
- head of product planning (P),
- head of sales (V) (essential for product ideas),
- as necessary, specialists and potentially corporate management (G).

If possible, do not delegate E, F, P, and V (maintain the overview and responsibility).

HOW: There are no general regulations, but *possibilities* include:

- Free discussion based on the evaluation lists. Basing decisions solely on rating lists is insufficient. However, the ten highest rated ideas should be the most useful. One potential problem is the personal assertiveness of individual participants.

- Consulting of experts. One potential problem is that experts are not infallible and often one-sided. Check that their arguments have a solid foundation.
- Further action on the idea may be postponed until later in the planning phase so that further information can be obtained (e.g., extended evaluation methods).

2.2 Design Phase

Significance: The design phase starts with the cleared project and ends with a primarily functional product with predictable and justifiable production costs and a suitable market. The project is terminated if it no longer appears to make sense in terms of function, price, or market.

9 Start project

WHO: Head of product planning, together with head of product development.

WHAT: In particular, the following must be *determined*:

- the responsible project leader and the members of the project group (temporary team formation),
- the name of the project (short but meaningful),
- a reference number for the project (project number); this number must be

  - clearly arranged (no abstract sequence of digits) and
  - sortable (for filing cabinet or computer data base).
  - For example, the number G-21.096/2 has the following significance:

    G—project group (e.g., gardening equipment),
    21—starting year,
    096—serial number of projects within the project group,
    2—serial number of reports or similar documents for the project.

10 Create schedule

WHO: Project leader (see Point 9).

WHAT: *Recommendations* for this are as follows (see Fig. 4 for an example):

- Write down all important procedures.
- Arrange the order of procedures in a logical sequence.
- Estimate the time required for individual procedures (with safety margins).
- Derive deadlines from this and compare them with the target (see Point 8).

HOW: The *methods* for this may include the following:

- *Milestone plan* (bar chart, Gantt chart, Excel table): In many cases, this method is sufficient. It is clear and easy to change (e.g., pegboard).

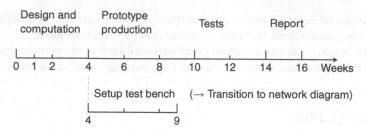

Fig. 4 Example of a project schedule

- *Network diagram:* This method reflects the activities and dependencies in the project, allows optimal project scheduling, and can be edited on a computer.

11 Estimate costs

WHO: Project leader, together with head of product development (or similar individual).

WHAT: The following must be estimated:

- *personnel costs:* number of employees × salary or wages × overhead factor × time required;
- *material costs:* material, test benches, etc.; and
- *other costs:* with safety margins (compare this with the cost framework in Point 8).

12 Ensure monitoring

WHO: Head of product planning, together with project leader.

WHAT: Any (self-evident) supervision by the project leader alone is not enough. A superior body must control important deadlines (milestones).

HOW: *Examples* of this include the following:

- recurring meetings,
- status information by e-mail; otherwise, follow-up.

13 Ensure functional optimization

WHAT: The order of the following Points 14 to 18 is not a mandatory sequence but rather an example of various activities at the processing level. They can run in parallel, be replaced by others, omitted, added, etc.

WHY: Functional optimization means first designing the product to be fully functional, taking into account all essential requirements, before other aspects (e.g., production-related, cost-related, aesthetic) are used in the next phase for overall optimization.

14   Conduct market research

WHO: Individuals in sales, distribution, marketing, or similar functions.

WHAT: *Examples* of this include:

- information about competing products,
- updated patent search,
- determination of possible prices and sales volumes (if not already done in Point 4).

HOW: There are various ways to complete this task, from random sample surveys of customers to extensive investigation activities.

15   Conduct development design

WHAT: Central design activity.

HOW: *Examples* of this include:

- application of design methodology, design theory, creativity methods, etc. and
- computer-aided simulation.

16   Conduct short, comparative costing

WHY: A rough calculation of the manufacturing costs already in the stage of the design drawings is needed.

HOW: There are *two types of calculations* that can be used:

a) short calculation (e.g., according to "VDI 2225 Part 1" 1997), which is especially useful for comparing different designs;

b) precalculation based on complete drawings (although this is often too complex a method to be used at this stage).

WHO: Depends on the type of cost calculation:

To a)   product development (together with production),
To b)   production (together with costing).

17   Produce prototypes

WHAT: There are *two types* of prototypes:

- prototypes for the company's own tests and
- demonstration and test prototypes for customers, especially bulk buyers.

HOW: This involves mostly individual production (e.g., by means of additive manufacturing). The decisive factor is *functional* conformity with the later series product.

18   Conduct tests

WHY: Without sufficient testing, a meaningful evaluation of a product or process is rarely possible. (An exception is individual production, where acceptance tests take place after the production of the product.)

HOW: The *criteria for test conditions* are as follows (they may partly contradict each other):

- as similar as possible to practical conditions,
- clear and reproducible,
- justifiable effort.

The initial aim is to change only one influencing variable per test series in order to make the influence of that variable visible. A disadvantage is the high expenditure of time. For this reason, there are methods for limiting the experimental effort in a clever way with the mathematical resolution of the influences of various parameters (multifactor analysis, Taguchi method, design of experiments, etc.). If possible, an attempt should be made to trace a problem back to *dimensionless parameters* or to *model laws*; however, this requires knowledge of all significant influencing variables.

Types of experiments include the following:

- model tests (e.g., friction tests with two rollers),
- component tests (e.g., gear wheel in gear test bench),
- product tests (e.g., truck transmission on a test bench), and
- field tests (e.g., truck in the field).

In this order, practical relevance and costs increase while reproducibility decreases.

The *structure of a test report* can be as follows:

a) *Head:* Includes a meaningful, short title and a code number for storing, filing, and/or classification (see Points 9 and 38).
b) *Task:* Includes the purpose and objective of the experiments in a few words; otherwise, the report will be incomprehensible later (this is a frequent error).
c) *Summary of results:* Belongs on the front page for quick information.
d) *Experimental setup:* Includes a principle sketch, important equipment, and brief notes.
e) *Experimental procedure:* Includes an explanation of the procedure; otherwise, it will be difficult to estimate the validity of the experiment later.
f) *Test results:* Includes the results in numerical values and, especially helpful, in diagrams (which are quicker to understand). The results include the following:

- comment on the results,
- error estimation, and
- area of validity.

Points 15 to 18 are usually run several times (iterative) because the test or calculation results lead to design improvements and thus to new patterns and tests. Meanwhile, the project leader and the planning management monitor deadlines and costs; if these are exceeded, measures must be decided upon and initiated with the responsible bodies.

19 Evaluate results

WHO: Project leader.

WHAT: This involves evaluating the reports (market, calculation, tests) for the following:

- completeness,
- logical coherence (have it completed if necessary), and
- compliance with the requirements.

20 Create report

WHO: Project leader.

WHAT: Brief summary of the main findings with a proposal for the continuation, modification, or termination of the project.

HOW: In principle, similar to the experimental report (see Point 18), as it includes:

- the project name and reference number (see Point 9),
- the task definition (essence in short),
- a summary of the results with a proposal for further action,
- a short explanation of important results, and
- individual reports as attachments.

21 Arrange meeting

WHO: Head of product planning (as competent body).

HOW: *Recommendations* are as follows:

- arrange an appointment with the participants (by telephone, via Doodle, or similar) and
- provide a written invitation with an agenda and documents at least five working days before the meeting.

22 Make development decision

WHO: Product meeting, consisting of:

- head of product development (E),
- head of production (F),
- head of materials management (M),

- head of product planning (P),
- head of quality management (Q),
- head of sales (V),
- project leader (L), and
- where appropriate, corporate management (G), head of finance, specialists, or directly appointed competent staff with decision-making powers.

WHAT: This involves making a decision about the project's

- continuation (release for elaboration),
- modification (return to the design phase) or
- termination (give up; documentation, see Point 38).

If the project is to be continued, a specified task will be defined for the elaboration phase. This includes:

- the functional requirements (revision of the requirements list),
- a time and cost framework (revision of the product costs if necessary), and
- a personnel framework (if necessary).

HOW: Similar to Point 8.

2.3   Elaboration Phase

Significance: The elaboration phase starts with the project being released for elaboration to a product or process that fully meets the task at hand and is ready for, for example, series production; otherwise, the project is terminated. While the design phase generally focuses more on functionality (functional optimization), the elaboration phase focuses on the overall optimization of the product in terms of, for example, manufacturing, economic, ergonomic, and environmental aspects.

23   Create schedule

WHO: Project leader.

HOW: This is frequently done as a network diagram; otherwise, as a milestone plan (see Point 10).

WHY: This is important, for example, to procure operating resources or sample external parts (long duration, critical path in the network diagram). Example: Production of injection molds; here, ensuring the injection conditions and dimensional accuracy are important.

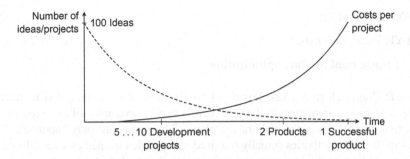

Fig. 5 Time course of ideas and costs (qualitative)

24   Create cost plan

WHO: Project leader, together with head of product development and, if necessary, other division heads.

HOW: The basis for this is the framework developed in Point 21. Here, more precise estimations (with in-house cost rates) can be made for the following:

- personnel costs (number × salary or wages × overhead rate × time);
- material costs (semifinished products, purchased parts, fabrications);
- equipment costs (new production, fixed and running costs for existing equipment); and
- other costs (orders to third parties, e.g., patent/market research; do not forget safety margins!).

WHY: Costs increase rapidly over time in the elaboration phase (see Fig. 5), which is why it is important to maintain an overview of the costs; if the actual costs overrun the predicted costs, a warning should go out and a product meeting or a similar problem-solving action should be initiated.

25   Initiate processing

WHO: Project leader.

WHAT: All affected parties (they may be scattered across the entire company) must be informed in a timely and clear manner.

HOW: The process depends on the company's organization. A general principle is that too much information is better than too little. Furthermore, personal consultations (where explanations and queries are possible) are better than subjecting those involved to "mail overload." It is helpful to attach a time schedule.

26   Ensure monitoring

WHO: Head of product planning.

HOW: See Point 12.

WHAT: Dates and costs.

27   Ensure total product optimization

WHAT: Points 28 to 31, like Points 14 to 18, are not to be regarded as necessarily logical or comprehensive but are merely typical examples of the wide range of activities possible in the elaboration phase. The points are only "addresses" of the activities. The activities actually required must be determined on a case-by-case basis depending on the company and project (see Point 23).

28   Conduct detailed design

WHAT: Elaboration, preferably in close cooperation with production, work preparation, quality management, and other affected parties (e.g., materials management, sales, and suppliers).

29   Produce preliminary, prototype, and pilot series

WHO: Production (not prototyping), together with quality management.

HOW: Production occurs on a trial basis with the original tools under series conditions (as much as is feasible).

WHY: The *reason* this is important it that it allows for the following:

* testing of the production equipment,
* setting up of machines,
* testing of the assembly process,
* acceptance of the products through quality assurance, and
* testing of the parts that are produced, where appropriate.

30   Make precalculation

WHO: Production.

HOW: As exactly as possible (must be checked and potentially revised after identifying necessary changes; see Point 32).

31   Conduct market and customer tests

WHO: Sales, in cooperation with product development.

WHAT: Presentation of the new product on a trial basis to some selected customers; if necessary, delivery of test prototypes to special customers (although caution is warranted here).

WHY: Gathering opinions and, if necessary, experience from the point of view of the user or customer.

32   Initiate drawing modifications

WHO: Project leader.

WHAT: Based on the experience gained, new, final drawings are created, or the final drawing status is determined.

HOW: During the elaboration phase, all modifications to drawings are centrally registered (usually, there are quite a few) and recorded in one place (e.g., in CAD or as a set of drawings). Otherwise, there will be no overview of the latest valid drawing status. Note: It is important to define who is allowed to modify drawings (otherwise, there may be chaos).

WHY: Due to a possible large number of modifications, discrepancies can easily occur; therefore, "new" drawings should always be created (and fixed) and only the final version should be released for access by other parties.

33   Evaluate results

WHO: Project leader.

WHAT: See Point 19.

34   Create report

WHO: Project leader.

WHAT: Similar to the test report; see Points 18 and 20.

35   Start release

WHO: Project leader.

WHAT: On the release form (to which all decision-related documents such as draw-ings and reports are attached), each of the decision makers (see Point 37) confirms with his signature that:

- he *agrees* to the release of the product for series production,
- the release will be granted only under certain (clearly specified) *conditions*, or
- he does *not* agree with the release (and states the reasons).

WHY: The form forces decision makers to deal with the project and its status in a timely manner and to record their approval or objections in writing. The further along the development is in the sense of "simultaneous engineering," the easier the release will be.

HOW: For an example of a release form, see Fig. 8.

36   Arrange meeting

WHO: Head of product planning.

HOW: See Point 21.

37   Make production decision

WHO: The *decision makers* include the following:

- head of product development (E),
- head of production (F),
- corporate management (G),
- head of materials management (M),
- head of product planning (P),
- head of quality management (Q),
- head of sales (V),
- head of finance,
- specialists as required, and
- project leader (L).

Do not delegate this task, as there are high costs if wrong decision is made.

WHAT: *Decisions* are made about

- the start of production,
- the improvement of the product (reworking), or
- the termination of the project.

HOW: If the project has been well prepared (see Points 32 and 33), the decision can be made relatively easily; otherwise, if necessary, other decision-making tools can be used (e.g., point evaluation, utility analysis, economic indicators).

38   Document project

WHAT: Systematically ordered, clear compilation of all documents relating to a completed project, including all rejected solution approaches and the decision documents.

WHO: Project leader.

HOW: The documents should be securely kept (to avoid leaks) but with clear access routes for leading personell in product development.

WHY: The *reason* this is important it that it allows for:

- collection of know-how,
- suggestions for new ideas for future similar projects, and
- decision support for such projects.

This is an extremely important activity that should not be neglected!

WHEN: This step should be done immediately after the completion of the project (otherwise, there is a risk of being left behind).

3 Creation of the Company's Own Schedule

3.1 Methodical Approach

Significance: Each company should develop and test its own schedule for product development projects tailored to its own needs and circumstances. The following steps and check questions are useful in this process:

1) *Activities:* Determine which development activities should usually be carried out and in which order.
2) *First draft of a flow chart:* Describe the process, as determined in Step 1, as a scheme or diagram.
3) *Review:* Review the process from Step 2 by asking the following questions:

   - Are *all* necessary activities included? (Add any that are missing.)
   - Are certain activities *superfluous*? (Delete any that are unnecessary.)
   - Is the process logically *clear* and as *simple* as possible? (Improve the process as necessary.)
   - Who has to *decide* on what and at which points? (A decision should be made at least three times and specified exactly.)
   - Who *monitors* which deadlines and costs? What happens if they are exceeded? (This should be recorded exactly.)

4) *Improved flow chart draft:* Design a new schedule according to the results of Step 3.
5) *Discussion:* Discuss the plan from Step 4 with *superiors, employees,* and *colleagues* at the same organizational level, as they will all have different experiences and perspectives.
6) *Preliminary flow chart:* Draw up a preliminary procedural plan and introduce it for a certain period of time (e.g., one year) to gain experience.
7) *Exchange of experiences:* After the deadline, discuss the experiences with all concerned (see Step 4).
8) *Final schedule:* Draw up and introduce a final plan based on the results of Step 7. In the event of major deviations from the original plan, provide for a further trial period or date for reviewing the new plan if necessary.

3.2 Drawing Modification, Deviation Approval, and Release

Significance: The three terms are closely related in terms of content and are therefore dealt with together here. Drawing modifications and deviation approvals are especially important during ongoing product production, that is, after product development is completed.

Drawing modification: Planned, permanent modification of the production documents; it requires an appropriate organizational framework.

Deviation approval: One-time permission for a nonfunctional feature of the product to deviate from the drawing. The drawing itself should not change. The deviation approval is related to pronounced emergency and must be an exception. Caution: This may incur overhead cost errors, so documentation is important in terms of product liability.

Release: Approval of a newly designed or modified product or part for (series) production. It is usually dependent on the written approval of the responsible decision maker.

In the case of small changes, the release procedure may become too complex. For this reason, criteria should be defined for when a formal release can be omitted or when it must be carried out in any case. A simple example is that there are no releases for individually produced parts, and for series parts, a new release always occurs after a drawing modification.

Organizational procedure: The organization of drawing modifications must be adapted to the respective operational conditions (production structure, organizational structure, etc.). In principle, the functions shown in Fig. 6 should be available. The principle diagram is to go through twice when changes are made. A release approximately corresponds to Steps 1 to 6. The remaining steps concern the operational implementation.

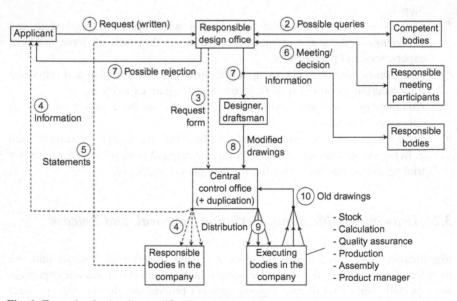

Fig. 6 Example of a drawing modification system

Organizational specifications of a drawing modification system: The following questions can help clarify and define the organizational process:

a) Who is *responsible* for modifications (responsible unit)?
b) Who has to *comment* on requested modifications (responsible units within or outside the organization, e.g., for joint projects)?
c) Who must *participate* in a modification decision?
d) When does a modification require official *approval* (criteria)?
e) Who must be *informed* before a modification is implemented (responsible units)?
f) Who controls and monitors the implementation (responsible *central* unit)?
g) When is a modification made in a *run* production (e.g., run-through of a series)?
h) Who *distributes* the modified drawings and collects the old ones?
i) *When* are the old drawings withdrawn (e.g., immediately with the issue of the new drawings or overlapping)?
j) Who reports the *completion* of a modification to the central unit and how?
k) How and where can the status be found *prior* to the documented modification (e.g., file organization)?
l) Who may/can/must have *access* to the IT system and in what form?
m) How can access by authorized or unauthorized parties be *enabled* or *prevented*?

Forms: Figs. 7 and 8 each show a shortened example of a drawing modification/deviation approval and a release form. Computer-supported documentation of changes is highly advised.

○ Drawing modification	○ Deviation approval	Product group:	No.:	Distribution list:
Applicant:			Date:	
Subject: (Product, part)			Case no.:	
Modification/deviation:				

Explanation:

Statement:						
Name	Yes	Yes* if	No*	Does not concern	Date	Signature
*Condition/reason						

Fig. 7 Example of an application form for drawing modification and deviation approval

RELEASE FORM	Product group:	No.:	Distribution list:
Applicant:		Date:	
Subject: (Product, part)		Case no.:	
Calculated quantity:	Intended release date:		
Explanations: (Documents are attached)			

Release note:					
Area	Yes	Yes* if	No*	Date	Signature
*Condition/reason					

Fig. 8 Example of a release form (in "warning color")

References

Jorden, W., & Weiberg, H. (1977). Systematische Entwicklung einer Baureihe von Spreizbüchsen-Spanndornen. *Konstruktion, 29*(2), 55–61.

VDI Guideline 2221. (1993). *Systematic approach to the development and design of technical systems and products.* Berlin: Beuth (in German).

VDI Guideline 2225 Part 1. (1997). *Design engineering methodics—Engineering design at optimum cost—Simplified calculation of costs.* Berlin: Beuth (in German).

Methodical Procedure in a Product Development Project

1 Development of a Design Methodology

General: There are a number of different design methodologies in existence today. They differ from one another through, among other things, the subject area (industry) or the personal viewpoint of the originators (see Pahl et al., 2007). The framework for these methodologies is provided by "VDI 2221" (1993).

Problem-solving process: Designing is a problem-solving process; consequently, the procedure underlying such a process must also be found here. The problem-solving process takes place in five steps (see Jorden, 1983):

1) *information* (problem),
2) *definition* (goal),
3) *creation* (ideas),
4) *evaluation* (criticism), and
5) *selection* (solution).

If necessary, the evaluation is followed by change and improvement, that is, a new round of creation, evaluation, and selection activities.

Design process: According to Jorden (1983), the activity of an individual designer or a small design group can be divided into (at least) *five stages* (see Table 1). Each of these stages may contain the five steps of the problem-solving process but does not have to; that is, steps can be skipped if doing so makes sense. Often, the following combination of steps is included:

1) production of *solution ideas* ("green phase"),
2) *evaluation* of the solution ideas ("red phase"), and
3) *selection* of the most suitable solution.

Differences from "VDI 2221" (1993): "VDI 2221" (1993) distinguishes *seven* stages instead of the five seen in Table 1; however, the basic scheme can still be the same as that in Table 1. The additional *modularization* found in "VDI 2221" (1993)

© The Author(s), under exclusive license to Springer Nature Switzerland AG 2021   61
J. Schlattmann and A. Seibel, *Structure and Organization of Product Development Projects*, https://doi.org/10.1007/978-3-030-81046-7_5

Table 1 The five stages of design. Based on Jorden (1983)

Stages	Results of the stages
Task	Requirements list
Functions	Function list, function structure
Working principles	Principle solutions, principle sketches
Design elements	Modularization, full-scale sketches
Overall design	Drawings set, parts list, documentation

Table 2 The three phases of product development. Based on Jorden (1983)

Phases	Results of the phases
Planning	Task (project)
Designing	Design drawings, functional models (functionally optimized)
Elaborating	Production documents, prototype (overall optimized)

occurs at various points in the design process. In Table 1, for example, it is included in the "design elements" stage. The *documentation* stage is extremely important but is not part of the design process; it is the responsibility of those at the control level (project leader).

Development process: According to Jorden (1983), the activities of product development are divided into (at least) *three phases* (e.g., planning, designing, elaborating), each of which is *followed by a decision*. A distinction is to be made between the processing, controlling, and deciding levels. The three phases of the development process and the correlating results are shown schematically in Table 2.

Relationship between the phases and stages: The *five design stages* (see Table 1) appear in *each phase* but with different levels of emphasis; in addition, there are other activities that may take place in each phase (e.g., tests, market studies, prototype design). For this reason, phases and stages cannot generally be directly matched to each other, as is often attempted in the literature. The mixing of "phases" and "stages" leads to the construction of complicated and confusing methods—as they are sometimes described—, which is the main reason that design methodology has been difficult to apply in practice before now. The connection between phases and stages can be illustrated as in Fig. 1. However, the scheme presented here is neither complete nor mandatory in the design process.

Relative importance of the stages within the phases: The five design stages appear in all three phases of product development. The direct connection between phases and stages is determined by the shift in emphasis among the stages in each phase, which can be characterized as seen in Table 3 (there are also considerable differences in emphasis depending on the task); see Jorden (1983):

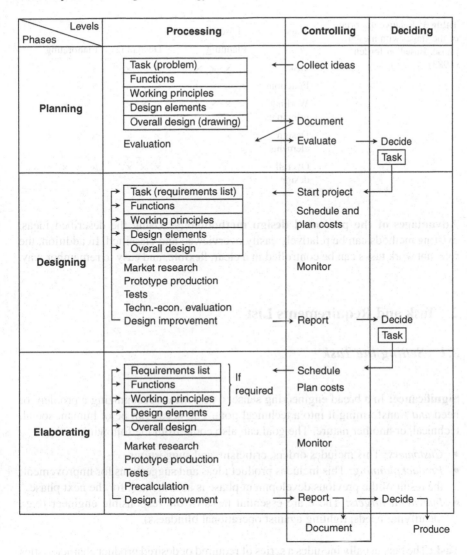

Fig. 1 Relationship between the phases and stages. Based on Jorden (1983)

- In the *planning phase*, the emphasis is primarily on the task and the functions, and the considerations can extend to the overall design.
- In the *design phase*, all five stages have equal importance.
- In the *elaboration phase*, the emphasis is on the design elements and the overall design; the task, functions, and working principles can also be referred to if necessary.

Table 3 Relative importance of the stages within each phase. Based on Jorden (1983)

Stages	Phases		
	Planning	Designing	Elaborating
Task	× × × × ×	× × ×	×
Functions	× × × ×	× × ×	× ×
Working principles	× × ×	× × ×	× × ×
Design elements	× ×	× × ×	× × × ×
Overall design	×	× × ×	× × × × ×

Advantages of the presented design methodology: Using the described ideas, existing methods can be relatively easily overviewed and classified. In addition, the relevant work tasks can be controlled in a clear, flexible, and easy to remember way.

2   Task and Requirements List

2.1   Setting the Task

Significance: In a broad engineering sense, a task means recognizing a problem or need *and* transforming it into a technical goal. The needs can be of human, social, technical, or another nature. The goal can also come from the following sources:

- *Customers:* This includes orders, criticism, and suggestions.
- *Product planning:* This includes product ideas and suggestions for improvement; the result of the previous development phase is also the task for the next phase.
- *Individual targets:* These are essential tasks of the responsible engineer (e.g., identifying needs, fighting against operational blindness).

Task: The task usually includes a series of required or desired product characteristics or main functions (usually formulated in writing). These specifications cannot usually be immediately or directly used for the design as they are often incomplete and/or unclear.

2.2   Clarifying the Task

Significance: The task definition must always be checked and supplemented or corrected as needed. Incorrect or unknown requirements or boundary conditions mean that the design has to be changed later on, usually with considerable effort, or it may even become completely unusable.

Basic questions: The designer should always ask himself the following questions:

- Which *main function* (which purpose) should the design fulfill?
- Which properties *must, should, may* it have, and which properties *must* it *not* have?

Clarifying questions: The following checklist can be used to determine the shortcomings of the existing information:

- What are *demands*, what are *wishes*? ("Must" and "may not" are demands.)
- Are the requirements *clear, realistic, truly necessary,* or *contradictory*?
- What *additional* information is required (e.g., recyclability, manufacturing)?
- *Who* sets the requirements (customer or manufacturer—field service, development, planning, production)?

These questions must be clarified immediately with the person making the request, which can be done:

- *personally:* allows for dialogue and visual inspection (this is the most favorable option but is often costly);
- *by telephone:* allows for dialogue, is fast, eliminates problems of distance; or
- *in written form:* it is a lengthy process and carries the risk of misunderstanding, but it allows information to be clearly documented.

The outcomes of a conversation or telephone call should always be recorded and confirmed in writing.

Additional information: In general, the designer must add further information on his own initiative based on his knowledge and experience. This information may be:

- an evaluation of customer complaints (product weaknesses);
- an evaluation of customer questions (general wishes, claims);
- about standards (ISO, company standards);
- about repeat parts (rational production, interchangeability);
- about the state of the art (competitiveness);
- about future technical-economic development (future-proof products); or
- about quality assurance (documentation, product liability).

2.3 Requirements List

Significance: The requirements list contains all information necessary for the execution of the task in a clearly arranged form. It is compiled by the design engineer and is the reference for the design result (and thus also the basis for evaluating the solutions). It is a document (like the drawings) and therefore subject to change management.

Content: The requirements list includes:

- *Demands:* These must be fulfilled; otherwise, changes are necessary. A distinction is made between

  - *fixed demands* (e.g., a rack height of 3500 mm) and
  - *limited demands* (e.g., a rack height of at least 3500 mm).

- *Wishes:* These should be fulfilled as far as possible (they may be discretionary or decisional and are often dependent on costs). It may be necessary to prioritize among wishes.

Demands and wishes should be clearly marked (e.g., D and W). Both should indicate, if possible, information on:

- *quality* (verbally formulated) and
- *quantity* (numerical values, limit deviations).

A wish becomes a demand through numerical values; otherwise, it is called a "reference value," denoted with the words "if possible," etc.

Structure: The requirements can be structured in a table with the following information:

- *project name*, project number, and client;
- *demand or wish*, potentially *goal* (e.g., long-term goal); it should be marked (e.g., D, W, G);
- *origin* of each requirement (e.g., the company itself or customer);
- requirements broken down by *groups of characteristics* (see below);
- indication of the *state of change* (index, date, change, previous state, cause).

Outline: The requirements list can be structured or checked for completeness using the information in Table 4. The outline can be shortened, summarized, or extended as required.

Table 4 Example outline for a requirements list

Functional requirements	Manufacturing requirements	User requirements
Geometry	Costs	Usage
Kinematics	Schedule	Safety
Forces	Series size	Ergonomics
Energy (conversion)	Assembly	Environment
Substance (conversion)	Quality assurance	Maintenance
Information (conversion)	Transport	Recycling

Example: Fig. 2 shows an example of a requirements list for a testing machine that simulates the idling behavior of multidisc friction clutches.

		REQUIREMENTS LIST Friction clutch idling test bench	No.: VT 95 014 Date: 03/28/2021 Page: 1	
		Client: Research Association Drive Technology		
No.	D/W	Requirements		Chg.
1 1.1 1.2 1.3 1.4 1.5	 D D D D W	**Geometry** Variable number of discs ≥ 6 Largest outer diameter of inner discs $\geq 200\,mm$ Smallest outer diameter $\leq 60\,mm$ Inner and outer discs can be offset against each other: - radially up to $\geq 0.3\,mm$ - angularly up to $\geq 0.001\,rad$ (0.06°) - wobbling of a clutch side up to $\geq 0.001\,rad$ Possibility to tilt the axis up to 15° against the horizontal		
2 2.1 2.2 2.3 2.4 2.5	 D D D D W	**Kinematics** Counter rotation of both disk packs or one pack stationary Differential speed at circumference $\geq 70\,m/s$ Speed continuously adjustable within ranges Torsional vibrations must be superimposed on the drive, corresponding to an impact factor of $S = 1.2$ for diesel engines. Speed range up to around 0 rpm		①
3 3.1	 W	**Energy** Energy-saving test bench concept should be aimed for.		
4 4.1 4.2	 D D	**Substance** Oil quantity continuously adjustable up to approx. $\leq 10\ 1/1000\,cm^2$ friction face Wet and dry running of the multi-disc clutch		
5 5.1	 D	**Schedule** Fast production of the components; completion ≤ 7 months		
6 6.1	 W	**Costs** Use of commercially available components should be aimed for.		
		Note: In this list, the requirements are not identified by origin as they all originate from the client.		
①	2.2	03/28/2021	Differential speed 70 m/s (previously 50 m/s) according to meeting from 03/25/2021	
Ind.	No.	Date	Change	

Fig. 2 Example of a requirements list for a friction clutch test bench (shortened)

2.4   Guiding Rules for Creating Requirements Lists

1 Requirements list

In principle, a requirements list must be created for each design task or the existing list must be checked and supplemented (overall task, subtask); this should also be done for each development phase (in particular, "designing" and "elaborating").

2 Task clarification

Almost every task is initially incomplete. Therefore, information must be checked and clarified by making further inquiries. Particular attention must be paid to whether demands or wishes are truly necessary or are too narrowly defined (out of caution, unconscious limitations, etc.). These questions and the outline proposal (see above for both) can help in obtaining the most complete information possible.

3 Outline of requirements

The requirements should be clearly outlined according to groups of characteristics (see Table 4) and their origin (e.g., the company itself or customer) should be noted. Within these groups, it is advisable to list the damands first and then prioritize the wishes (i.e., to list the most important ones first).

4 Forms

Simple and clear forms (similar to that seen in Fig. 2) should be used for the requirements list and standardized within the company. All requirements lists within a company should follow the same structure.

> **5 Changes**
>
> Requirements lists are documents. They must therefore be managed and stored centrally and are subject to change management. In principle, changes are permitted only with the agreement of the ordering party (the party making the request). Corresponding meeting results must be recorded and delivered, as they are part of the documentation.

2.5 Abstraction to Identify the Core of the Problem

Significance: Even after being clarified, the requirements list is often still too narrowly focused on a specific problem solution, as it is oriented—more or less unconsciously—toward solutions to problems from the past. Furthermore, the true core of the problem often emerges only during an intensive examination of the task. Therefore, this topic is discussed here, although it truly belongs at the beginning of this section. Often, the actual core of the task can be worked out on the basis of the previous requirements list by omitting everything unimportant, which, under certain circumstances, leads to novel solutions. It often contains the overall function of the product and thus leads directly to Sect. 3.

Example: The following two example tasks illustrate the importance of abstraction (see Pahl et al., 2007):

- Develop a garage door.
- Develop a garage lock that protects against theft and weather and prevents others from seeing inside the garage.

Procedure: Based on the requirements list, the following steps are recommended (see Pahl et al., 2007):

1) Omit wishes (mentally).
2) Omit insignificant demands (mentally).
3) Reduce quantitative statements to essential qualitative ones.
4) Formulate the core problem in a solution-neutral way ("What is truly important?").

3 Functions

3.1 General Information

Significance: A function is the general relationship (described abstractly) between the input, internal, and output variables of a technical system for the fulfillment of a task ("VDI 2222 Part 1" 1997).

Formulation: A function should be formulated as simply as possible and as abstractly as necessary, preferably as a noun plus a verb. The abstraction should help to avoid prefixed problem solutions and open up room for other possible solutions (see "creativity methods").

3.2 Overall Function

Significance: The overall function is the abstracted main task of the system; under certain circumstances, several overall functions may result from the task.

Representation: This is done as a "black box," that is, as a change in the properties of the three flows between input (I) and output (O):

- *substance,*
- *energy,*
- *information.*

One of these is usually addressed as the main or most important flow. An example of an overall function with input and output flows is shown in Fig. 3.

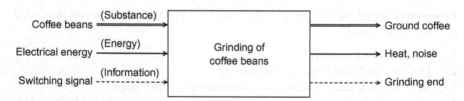

Fig. 3 Example of the overall function of a coffee grinder with input and output flows

Table 5 Examples of the general functions "conduct," "store," and "convert"

	Conduct	Store	Convert
Substance	Pipeline, conveyor belt	Shelf, container, tank	Combustion, deep drawing
Energy	Cable, compressed air hose	Battery, compressed air tank	Motor, jackhammer
Information	Light (conductor), data bus	CD, chip, flash drive	Typewriter, printing

Table 6 Examples of the general function "connect/separate"

Connect/separate	Substance	Energy	Information
Substance	Welding, mixing	Melting, pressing	Embossing, light barrier
Energy	–	Power steering	Sensor button
Information	–	–	PC

3.3 General Functions

Significance: Processes in technical systems can generally be divided into four categories:

- *conduct:* change in *location* (of a substance, energy, or information),
- *store:* change in *time* (related to a substance, energy, or information),
- *convert:* change in *appearance* (of a substance, energy, or information),
- *connect/separate:* *logical function*, examples of which include mixing, interrupting, or adding energy to affect a substance (e.g., pressing).

These general functions represent the highest level of abstraction. They can help in analyzing and understanding processes in technical systems. However, the extent to which their use is logical depends on the individual case. In many cases, it is better to use somewhat more concrete function formulations that are specific to the application (e.g., "actuate switch" instead of "connect information with energy"). Table 5 shows some examples of the general functions "conduct," "store," and "convert," and Table 6 shows examples of the general function "connect/separate."

3.4 Subfunctions

Significance: Subfunctions divide the overall function. There is no "mandatory" or "correct" division; the most appropriate division depends on the problem.

Table 7 Example of the function list of a coffee grinder

Input functions (IF)		Internal (system) functions (SF)		Output functions (OF)	
IF 1	Fill in beans	SF 1	Feed beans	OF 1	Dispense ground coffee
IF 2	Supply energy	SF 2	Grind beans	OF 2	Release heat
IF 3	Switch on energy	SF 3	Store ground coffee	OF 3	Emit noise
...		SF 4	Protect fingers	OF 4	Signal end of grinding
		...		...	

Procedure: The essential question is whether a technical problem solution already exists for the overall function. The following steps are useful in determining this:

a) *No suitable solution exists or is known:*

    1) Set up a black box.

    2) List recognizable subfunctions (see Table 7 for an example), separated according to their input, internal, and output. If necessary, first create a rough solution concept and then analyze (abstract) the subfunctions.

    3) Further break down the subfunctions for which no solution is available according to the same scheme.

b) *The existing solution only partially fulfills the functions* (so-called customized design, see Sect. 6.3):

    1) Set up recognizable subfunctions.

    2) Treat subfunctions without a suitable solution as described above under Point a 2, that is, determine group functions.

    3) Further break down the group functions without a suitable solution.

3.5 Function Structures

Significance: A function structure represents the logical relationship between the individual subfunctions in a circuit diagram. It can generally be determined from the input and output values of the subfunctions.

Representation: Arrows should indicate the flows (see Fig. 3) and the physical quantities (e.g., "ω" for angular velocity, "T" for torque). Complete and correct quantities should be given (clear identification of the flows). An example of a function structure is shown in Fig. 4.

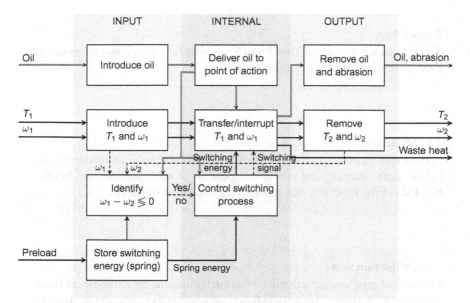

Fig. 4 Example of the function structure of a free-wheel clutch

Variation of the function structure: In most cases, only one function structure results from the subfunctions. Further structures and thus possible solutions result from the variation of the structure with the following possibilities:

- change the *sequence* (swap),
- consider *multiple arrangements* (add, i.e., further subdivide or combine),
- change the *circuit* (series or parallel),
- change the *feedback* (open-loop or closed-loop),
- move the *system boundary* (add or omit elements).

3.6 Guiding Rules for Functions and Function Structures

1 Formulation

Formulate functions as simply as possible and as abstractly as necessary (a noun plus a verb). Abstraction is a tool and not an end in itself. Too much abstraction is confusing, too little causes fixation on only one solution.

2 Black box

Set up a black box and important flows for substances, energy, and information. First, look for existing technical solutions.

3 Subfunctions

List the subfunctions based on the black box or existing solutions and separated by the input, internal, and output components of each subfunction. Further break down the functions that lack existing technical solutions.

4 Essential functions

Identify essential functions that decisively determine the design and mark them, for example, in color. Pursue these functions before others to focus on working principles and principle solutions.

5 Function structure

Combine the listed functions into an initial function structure as simply as possible; in the process, pay attention to providing clear, matching flows. Trace substance, energy, and information flows individually.

6 Structure variants

Systematically vary the initial function structure. After each variant has been set up, think about possible ways to implement it.

The mental crossing of boundaries between *abstract* and *concrete* can lead to new approaches to solutions.

7 Breakdown

Break down the critical points of the solution in the function structure as far as possible. A breakdown into *several* elements usually allows for a *better fulfillment* of the function but with *increased effort* ("differential design").

Conversely, combining several functions within *one* design element usually results in a simpler design but a *worse fulfillment* of the function ("integral design").

8 Application

Guiding Rules 1, 2, 3 and 4 should always be applied when the task requires this degree of abstraction, that is, when functions are not known or are insufficiently known or when new solutions are sought via new working principles. Furthermore, the functional analysis provides essential assistance in cost investigations (see "value analysis").

The function structure provides more detailed information about the inner logic of a design; it is recommended especially in cases involving more complicated relationships, as it can help them become more manageable. In many cases, however, it is not absolutely necessary (it costs time!); it is often sufficient to draw up a function list (see Table 7), select the essential functions, and pursue them with working principles and solutions.

There is neither a "correct" nor a "complete" function structure (at best, there is only a wrong one). However, it is also not possible to specify *all* function structures for a certain overall function.

4   Working Principles

4.1   General Information

Significance: A working principle enables the conversion of a (sub)function into a physical (chemical or other) event, initially without regard to its technical or economic feasibility. It is always based on a physical (or other) *effect* (see Table 8). The appropriate concreteness depends on the problem at hand. Often, the goal is to find as many working principles as possible for a function and thus identify new approaches to developing a solution.

Table 8 Physical (or other) effect, working principle, and solution principle

Term	Example	Description
Effect	Dry friction	Verbal (qualitative) or as a formula (quantitative); both should always be solution-neutral options
Working principle	Clamping	Combination of subfunction and effect enabling this subfunction; verbal or as a "black box"
Solution principle	Screw clamp	Combination of working principle and arrangement scheme (but without a design and dimensions)

Types of working principles: The following list gives suggestions and can be extended and refined as needed:

- mechanical (elastic, plastic; static, dynamic; form fit, force fit, material fit)
- electrical (ohmic, inductive, capacitive; magnetostrictive, piezoelectric, electromagnetic)
- hydraulic (hydrostatic, hydrodynamic).

Search for working principles: Creativity methods are especially useful for identifying working principles:

- *Intuitively emphasized methods:*

  - brainstorming,
  - variants of brainstorming (problem reformulation, brainwriting, 6–3–5 method),
  - image-based methods (bisociation method, analogy method),
  - synectics.

- *Logically emphasized methods:*

  - analysis of formulas,
  - morphological box,
  - targeted feature variation.

- Use of *design catalogs.* (This is not a creativity method!)

4.2 Developing Principle Solutions

Significance: A principle solution (concept) is a structure (arrangement) of working principles that together result in the physical (or other) event (see Table 8); the structure is similar to a function structure.

Procedure: Starting from the function structure, working principles (as many as possible) are reviewed to identify subfunctions, either

- *verbally:* a list of physical (or other) effects (clues for finding design elements);
- as a *working structure:* according to the function structure, with a specification of the physical (or other) quantities or effects; or
- as a *principle sketch:* a line representation with simple symbols (e.g., gear unit sketch).

Combining to achieve a principle solution (concept): Here, the search for compatible combinations of the working principles takes place (this is a critical point in the design process). The following methods are suitable for this (see Pahl et al., 2007):

- *"moving forward"* from the input variables,
- *"moving backward"* from the output variables,
- using a *morphological box.*

Proof of physical feasibility: A prerequisite for further processing is that the intended physical (or other) process is possible. The proof can be done by

- formulaic derivation and calculation,
- physical tests,
- computer simulations.

4.3 Designing with Design Catalogs

Significance: Design catalogs contain the most comprehensive possible collection of solutions for specific, recurring subtasks in a systematic and clearly structured form (information memory); see "VDI 2222 Part 2" (1982) or Roth (2000) and (2001) for more information.

Features: Design catalogs are:

- clearly and systematically structured (search help),
- as comprehensive as possible (complete, give an overview of an area),
- characterized by descriptive characteristics and examples (application help),
- coordinated with the design methodology (sections) if possible,
- preferably suitable for computer processing (algorithmization capability).

Types of design catalogs: According to their *content*, they can be divided into the following (although the boundaries between these are often fluid):

- *Object catalogs:* They contain technical objects that are necessary for the design process, and they are independent of the design task. Example: Rolling bearing catalog (types and sizes).
- *Operation catalogs:* They contain processes, process steps, or rules and their application conditions. Example: Rolling bearing calculation (e.g., service life).

- *Solution catalogs:* They contain the most complete range of solutions for a specific task. Example: Spindle bearing arrangements.

If the content of an area to be displayed is very large, catalog systems can also be created; these include the following:

- *overview catalogs* and
- *detailed catalogs.*

According to the *degree of generality*, catalogs can also be divided into:

- *General design catalogs:* These can be used with any product group and are therefore not company specific. They may be created, for example, by university institutes. Example: Tool clamping systems.
- *Company-specific design catalogs:*

  - These can be used to provide internal information for designers (and others) in a company. Here, a university institute can provide methodical help, for example. Example: Special clamping tools of a company.
  - These can also include offer catalogs that convey information to third parties (less important here).

General design catalogs: Each catalog generally consists of four parts:

- *Structure:* This part contains the essential aspects that logically divide the main part, that is, the actual content of the catalog. It should be free of contradictions and make it possible to check the completeness of the catalog. It functions like a *table of contents* in a book.
- *Main content:* This part is the actual *content* of the catalog. It contains objects, operations (rules), or solutions in the form of sketches, equations, and texts.
- *Access:* This part contains access features that mark important properties, thus making it easier to find suitable objects, operations, or solutions of the main part. It can be extended by the user with new access features. It is similar to the *subject index* of a book.
- *Appendix:* This part contains notes, explanations, examples, etc. It can be freely supplemented by the user.

Even if other terms are used in a catalogue, the first three parts are always present in some form.

Company-internal information catalogs: The result of the collection, preparation, and presentation of design solutions that have been developed in a company over time carries powerful intellectual potential. Its use saves time and costs not only in the design process by avoiding multiple developments and introducing younger designers to the know-how of others more quickly but, above all, in work preparation and production, because every *new* design also requires new work plans, fixtures, etc. However, the preparation of existing solutions usually fails because of the time required and the lack of tools available for systematization. These catalogs should be structured similarly to general design catalogs. Looking at good and bad examples is often also useful.

Application: In principle, using a catalog is possible in all design stages (from the task to the overall design). In general catalogs, the focus is usually on the working principles; company-specific catalogs are more focused on the design elements or the overall design.

Comprehensibility: The decisive factor in determining whether to use a catalog in practice is its quick comprehensibility. Essential for this are pictorial or symbolic representations that make the principles (or similar aspects) immediately clear. A catalog that is difficult for an average talented engineer to understand should not be used. Comprehensibility may be more important than the strict and abstract logic of the terms used.

Computer application: The systematic catalog structure allows information to be stored and processed electronically (e.g., searching on a computer screen). For example, catalogs can be expanded into information- or knowledge-based systems together with correspondingly arranged calculation documents, literature references, material data, etc.

5 Design Elements

Significance: The term "design elements" can cover two different areas, namely the *division* of a larger system (machine, plant) into individual elements (groups, components, modules) or the scaled design of the surfaces that are important for the given function, the so-called *"working surfaces."* The following section mainly considers the latter aspect, with which the actual design process begins in the narrower sense.

5.1 Working Surfaces

Significance: Working surfaces are the surfaces of a part that enable its functions. In most cases, these are surfaces through which contact with adjacent surfaces, media, or similar occurs.

Design elements: Design elements are the spatially and materially designed working surfaces. They can be represented by machine elements but are often only parts of them.

Classification in the design process:

1) Perform a *rough calculation* (or similar action) to determine the approximate dimensions of the working surfaces.
2) Search for *existing elements* (e.g., in a catalog). If unsuccessful or unsuitable, proceed to the next step.
3) Sketch the working surfaces *true to scale* according to the working principle (if necessary, this can be done not true to scale at first and possibly spatially).
4) Form the design elements by establishing the *material connection* between the working surfaces; correct these if necessary.
5) *Systematically vary* the working surfaces to find further possible solutions.
6) Identify *deficiencies* in the individual design sketches.
7) If possible, *improve* the design sketches accordingly; if necessary, make more precise recalculations and corrections based on the scale representation.
8) *Evaluate* the drafts according to the requirements list.
9) *Select* the most favorable drafts.

Example: The working surfaces of a gear result from calculations and are based on the conditions of those surfaces' connections to adjacent parts: tooth flanks, boreholes, keyways, and lateral contact surfaces (see Fig. 5). The design element results, in turn, from boundary conditions such as the material, series size, and manufacturing process.

Fig. 5 **a** Working surfaces of a gear wheel, **b** design as a casting

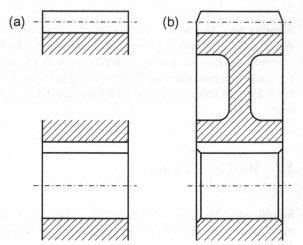

5.2 Modularization

Significance: Larger design projects, plants, etc. are first divided into individual feasible groups (modules). The structure often results from listing the functions or the function structure.

Modularization often leads to a *modular design* that allows the use of the same modules in different devices or design variants and thus to a more economical preparation of the offering, costing, production, storage, maintenance, and recycling of a product. The standardization of the interfaces of the modules (types of connections, connecting dimensions, etc.) is of the utmost importance here, not only internally within the company but also externally (e.g., for screws, rolling bearings, and plug-in cards).

Design stages and modularization: Basically, it is not possible to determine at which stage of the design process modularization should take place. It can happen after listing the functions or creating the function structure or after the development of principle solutions. For simple design tasks, it can even be omitted completely. For this reason, modularization is not listed here as a separate section—in contrast to "VDI 2221" (1993).

Examples:

- Modularization after the "functions" stage:

  - information processing devices (e.g., mechanical or electronic modules),
  - plant construction (division into large complexes).

- Modularization after the "working principles" stage:

  - mechanical designs (e.g., drive train, which includes the motor, intermediate flange, gearbox, and mounting).

Parallel processes: After modularization, each element (module, assembly) can be considered on its own in terms of design methodology. Finally, the modules are combined to form the overall design. In practice, this can lead to the overall design not being optimal because the cross information between the development groups is not sufficient or because of scheduling reasons, as one module is already being manufactured while the other is still under development and the module that has already been produced can then no longer be adapted.

6 Overall Design

6.1 General Information

Significance: The overall design is the true-to-scale representation of the required design task with all *necessary* details. What is "necessary" depends on the development phase (planning sketch, design drawing, or preparation for production). It is the basis for evaluations, decisions, and further procedures (e.g., elaboration or production).

Procedure: Due to the various influencing factors, there can be no binding steps in this process. However, the following procedure is recommended (see Schwarzkopf and Jorden 1984):

1) Select the *central point* of the design as the starting point (main point of action, main axis at the main point of action, etc.).
2) From there, design the *main working surfaces* according to the selected design elements; specify the zones to be kept free (e.g., for function and operation); extend the material to the working surfaces.
3) Proceed *outwards step by step*: Identify weaknesses in the current design and remedy them in the next step.
4) After a complete representation of the design is available, examine possible *disturbing influences* and *weaknesses* and revise the design. The combination of design elements that appear optimal on their own does not necessarily result in an optimal overall design; therefore, iteration is necessary.
5) In case of doubt, never change the same draft of the representation; *first save* it, and then edit it. The reason for this is the risk of errors when making changes; if necessary, keep the previous solution as a basis for further work.

6.2 Design

Significance: The design process involves the development and representation of the spatial and material form of the product, taking into account all of the requirements. It is the main design task, and it is here that the transition from "abstract" to "concrete" takes place. It is an extremely complex process and therefore is just as difficult as it is appealing.

Systematics of design: There are a variety of demands made upon any design (e.g., in the case of machine elements, strength and manufacturing as well as assembly processes are primary concerns); these demands range, for example, from ecological laws to aesthetic considerations. The demands also often conflict with each other and must therefore be weighed against each other to find an optimum compromise. More details on the systematics of design can be found, for example, in Pahl et al. (2007) and Ehrlenspiel and Meerkamm (2013).

6.3 Types of Design

Relation to design methodology: Depending on the nature of the task, not all stages of the design methodology need to be undertaken or equally intense (see Schwarzkopf and Jorden 1985). This is particularly dependent on the type of design required. A distinction is made between

- design types according to the *purpose of the order* and
- design types according to the *degree of novelty in the product.*

Design types according to the purpose of the order:

- *Development design:* Design (development) of series products due to the market situation and (generally) not for a specific customer; all design stages and development phases are passed through.
- *Offer design:* This is carried out on customer requests for the submission of an offer.
- *Commissioned design:* This is carried out on customer orders with (mostly) fixed tasks from the offer; there is deadline pressure, especially for the overall design in the elaboration phase.
- *Equipment design:* This is carried out for in-house production and is therefore usually assigned to the production area; the decisive factors here are functions and working principles, and the design is carried out with regard to safety and similar concerns.

Design types according to the degree of novelty in the product:

a) *Modular design:* All necessary components are available on call, and an assembly sketch or instruction is produced. Example: Gearbox according to catalog.
b) *Variant design:* The type of design elements remains the same, and a change is made mainly according to size and also according to the material or surface. Example: Gearbox with modified transmission ratio. A change in numbers, positions, and/or shapes leads to:
c) *Customized design:* The working principles remain essentially the same, but the design elements have to be partially redeveloped. Example: Gearbox with the output on top instead of on the side.
d) *New design:* The functions and working principles are new, even though existing experience (similar functions or principles) can usually be used. Example: Friction gear instead of tooth gear.

Table 9 Assignment of methodology stages to design types

Required stages	Design types			
	Modular design	Variant design	Customized design	New design
Task	×	×	×	×
Functions				× (×)
Working principles			(×)	×
Design elements		(×)	×	×
Overall design	(×)	×	×	×

× : reprocess; (×): change

The transitions between these design types are, almost always, sliding. The development of the modular elements necessary to reach Point a above is, in turn, a new design. Table 9 denotes the stages of the design methodology that are necessary for the different design types.

6.4 Methodology Plan

Significance: Before starting a design task, it should first be made clear which stages of the design methodology should or must be passed through and which should not. In general, those stages for which solutions or information are missing should be worked through; these stages generally depend on the degree of novelty involved in the task. This type of planning can logically be combined with personnel, time, and cost planning.

Procedure: Using Table 10, check the individual stages and select those to be processed. "Task" and "overall design" are generally included. Under "explanation," it can be noted what is important, what is missing, or why the stage will not be used. In the following columns, the responsible and additional employees are named. This is followed by the relevant dates and costs involved in the completion of the individual stages; they can be supplemented by actual data.

Responsibility: The head of design (group leader, project leader, etc.) is usually responsible for the entire methodology plan. The appropriate form of the plan for the company should be tested and determined.

Purpose: The methodology plan initially seems unappealing because it looks like bureaucracy and control. On closer inspection, however, the mistrust is unfounded because:

- The plan documents simply and clearly what has to be done.
- It refers (depending on the presentation) to suitable tools.

Table 10 Example of the structure of a methodology plan. Based on Schwarzkopf (1987)

Stages	Tools	To be processed		Responsibility	Employees	Schedule		Costs	
		Yes	Explanation			Target	Actual	Target	Actual
Task		×							
Functions									
Working principles									
Design elements									
Overall design		×							

- The need to determine personnel, deadlines, and costs cannot be avoided in rational design work.
- A target–performance comparison is necessary in an economic activity because it serves primarily as a basis for later, better planning (learning effect) and not as a means of control or even pressure.
- The plan exerts a "gentle compulsion"—which is especially necessary in the initial phase—to apply a design methodology. Without it, there is a risk of "falling asleep."

References

Ehrlenspiel, K., & Meerkamm, H. (2013). *Integrierte Produktentwicklung* (5th ed.). Munich: Hanser.

Jorden, W. (1983). Die Diskrepanz zwischen Konstruktionspraxis und Konstruktionsmethodik. In V. Hubka & M. M. Andreasen (Eds.), *Proceedings of the International Conference on Engineering Design* (Vol. 2, pp. 487–494). Zurich: Heurista.

Pahl, G., Beitz, W., Feldhusen, J., & Grote, K.-H. (2007). *Engineering Design. A Systematic Approach* (3rd ed.). London: Springer.

Roth, K. (2000). *Konstruieren mit Konstruktionskatalogen—Vol. I: Konstruktionslehre*. Berlin: Springer.

Roth, K. (2001). *Konstruieren mit Konstruktionskatalogen—Vol. II: Konstruktionskataloge*. Berlin: Springer.

Schwarzkopf, W., & Jorden, W. (1984). Die Gestaltung – Stiefkind der Konstruktionsmethodik? *Konstruktion, 36*(8), 299–304.

Schwarzkopf, W., & Jorden, W. (1985). Flexible Konstruktionsmethodik mit Hilfe eines Methodik-Baukastensystems. *Konstruktion, 37*(2), 73–77.

Schwarzkopf, W. (1987). Bildung eines flexiblen Systems für das konstruktionswissenschaftliche Methodenpotential unter Berücksichtigung der Anpassungsfähigkeit an praktische Anwendungsbedingungen. VDI-Fortschritt-Berichte, Series 1, Konstruktionstechnik/Maschinenelemente, No. 152. Dusseldorf: VDI.

VDI Guideline 2221. (1993). *Systematic approach to the development and design of technical systems and products*. Berlin: Beuth (in German).

VDI Guideline 2222 Part 1. (1997). *Design engineering methodics—Methodic development of solution principles*. Berlin: Beuth (in German).

VDI Guideline 2222 Part 2. (1982). *Design engineering methodics—Setting up and use of design catalogues*. Berlin: Beuth (in German).

Unfolding Creativity in a Product Development Project

1 Basics

1.1 Overview

Significance: "Creativity" means *creative power*, or the ability to produce something new, that is, something that—at least from the perspective of the creator—has not existed in this form before. It is the basis of all idea generation and thus a necessary prerequisite for innovative product development.

Forms of creativity: Of the numerous forms of creativity, Table 1 highlights some of the most notable. They are not clearly separable.

Characteristics of creativity: As illustrated in Table 1 and seen the literature, the following general statements are true (see Jorden, 1977):

- *Every* human being has creativity. This innate quality can be promoted and developed (to an individual degree; e.g., singing).
- Creativity involves both abilities and behavior. It is deeply rooted in the *unconscious*. Its promotion requires personality development.
- Creativity is important both for the *individual* (personality development, self-realization) and for the *general public* (as the basis and driver of all mental, social, technical, etc. development). It is vital for the existence of an industrial company.
- Creativity can be considered in terms of the *person* (e.g., child), the *result* (e.g., bicycle, law of nature), the *general public* (e.g., knowledge, progress), or the *creative process* (see the following section).

J. Schlattmann and A. Seibel, *Structure and Organization of Product Development Projects*, https://doi.org/10.1007/978-3-030-81046-7_6

Table 1 Forms of creativity. Based on Taylor (1959)

Term	Required skills	Result	Commitment, reactions	Examples
Naïve creativity	None	New only to the creator	Joy	Children's scribbles
Productive creativity	Certain abilities	New or of limited use/beauty	Personal commitment, pleasure	Hobbyists, hobby painter
Inventive creativity	Knowledge, abilities	Useful/valuable for general public	High effort, possibly to the point of ridiculousness	Karl Freiherr von Drais (bicycle)
Innovative creativity	Deep understanding	Significant change or insight	Use of personality, resistance	Johannes Kepler, impressionistic painting
Ingenious creativity	"Genius"	Breaks "the rules"	Fear, danger to life	Albert Einstein, Giordano Bruno

1.2 The Creative Process

Significance: The processes that take place during any creative endeavor can awaken a deeper understanding of the nature of creativity. Like all mental processes, they can be described only by means of images or analogous ideas; here, two extremely different examples are given (see Quiske et al., 1973).

Example 1 from history: One day, King Hieron II of Syracuse asked his friend, the mathematician Archimedes, to check whether a crown he had commissioned was actually made of solid gold or had been mixed with another, less valuable metal. The specific weight, that is, the weight per volume, of gold was known at the time, but Archimedes was unable to calculate the volume of the complicated crown mathematically. He thought very hard about the problem but was unable to find a solution. Then, one day when he was taking a bath, he observed that his body caused the water in the tub to rise and suddenly realized that all he had to do was place the crown in a cylindrical water vessel and calculate the volume of the displaced water using the difference in the water level before and after. He did this and found that the crown did not match the expected specific weight of gold; in other words, there had indeed been fraud.

Example 2 from behavioral research: A hungry chicken is behind a fenced-in space that is open on one side (Fig. 1a); outside the fence is some chicken feed. The chicken tries to fly over the fence or walk around it, but it turns back every time it loses sight of the grains: It is "blocked" and will starve to death without help. The generalization of the situation in Fig. 1b shows the problem (the unsolved task) like a wall between a person and the target that cannot be broken through directly, but only by a temporary distancing from the problem.

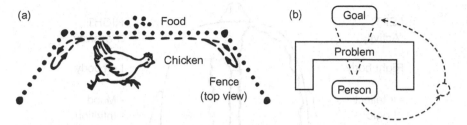

Fig. 1 Situation of a chicken blocked from food; **a** behind an open fence, **b** generalization of the situation. Based on Quiske et al. (1973)

The following insights into the creative process can be gained from the two examples and the literature.

Phases of the creative process: The creative process essentially takes place in *four phases* (see Poincaré, 1913):

1) *Preparation:* Intensive study of the problem; no solution.
2) *Incubation:* Waiting time, often frustrating; distancing from the problem.
3) *Inspiration:* Lightning-fast leap of knowledge from the unconscious or preconscious.
4) *Realization:* Review, evaluation, elaboration, and testing.

Activity and passivity: Something new always arises from the interaction of an *active* component (having the will to find a solution, making repeated advances toward a solution, showing tenacity—according to Edison, genius is only 1% inspiration and 99% perspiration) and a *passive* component (being able to let go, removing oneself from the problem, exercising patience, enduring frustration). The solution never comes from the conscious intellect; rather, it comes from the "unconscious" (or the intermediate instance, the "preconscious"). For example, on the journey home (passively), the solution suddenly occurs to us, whereas in the active phase, even with repeated attempts to find a solution, we do not succeed in finding the solution.

Polarity as the basis of human consciousness: The human mind can gain knowledge only by splitting into two opposite *poles*. For example, the term "positive numbers" is completely incomprehensible if there are also no "negative numbers." Both poles are mutually inseparable. However, people tend to overemphasize one pole and neglect or reject the other (see Fig. 2). This leads to one-sidedness and, ultimately, to the inability to make progress, as if someone wanted to use only the positive pole of a battery but not the negative pole. The essential thing is to see through the polarity and accept the right of the "negative" to exist.

Pattern formation: The human mind creates patterns (behavioral patterns, analogous to the subprograms of a computer) that control the behavior or the recognition of everyday processes or objects (such as a door) without us noticing those patterns

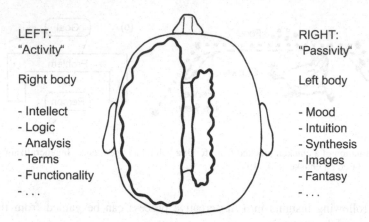

LEFT:
"Activity"

Right body

- Intellect
- Logic
- Analysis
- Terms
- Functionality
- ...

RIGHT:
"Passivity"

Left body

- Mood
- Intuition
- Synthesis
- Images
- Fantasy
- ...

Fig. 2 One-sidedness of "modern" humans. Based on Jorden (1988)

or bringing them specifically into our consciousness. Feelings of success increasingly lead to the formation of patterns. According to Jorden (1977), the following statements apply:

- In the case of a *problem, all suitable patterns* are first mobilized, that is, those that have previously led to success in similar situations (see the chicken example above). If the attempted solution is insufficient, frustration arises, and the process is blockaded. The blockade can be lifted only by the temporary distancing from the problem ("letting go"; see meditation, where one's will does not necessarily emerge from the unconscious).
- A *new* insight arises *unconsciously* and rapidly from the combination of different patterns that already exist but have not previously been considered together (see Archimedes example).
- Pattern formation is what makes human consciousness *viable* in the first place. It arranges and selects the flood of information rushing at us, letting in only what is needed penetrate our consciousness (even if we spot a new traffic sign on the way to work—we do not necessarily "see" it).
- Pattern formation means *routine formation* (similar to subroutines in a computer). Fixed patterns lead to habitual behavior and thus to an inability to engage in new activities (see Sect. 3). Too few patterns (a lack of knowledge) means a lack of breeding ground for creativity. Therefore, creativity always requires a sufficient number of flexible patterns.

Association: This means combining mental content in such a way that the presence of certain content in the consciousness causes or favors the presence of other content (stimulus word effect). This process happens unconsciously and continuously. Depending on the strength of the triggering stimulus, the consciousness is

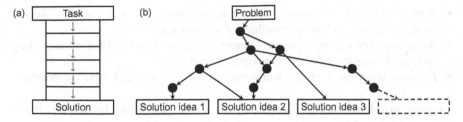

Fig. 3 **a** Logical (convergent) and **b** intuitive (divergent) thinking

"absent" for a shorter or longer period of time and then jumps back again (analogous to time-sharing with the computer). Chains of associations are the basis of our thinking.

Logical and intuitive thinking: Logical (convergent) thinking, which is currently overemphasized in our culture, leads step by step from a task (i.e., an already structured problem with a recognizable solution) to a solution (Fig. 3a), which can usually be evaluated as right or wrong. Intuitive (divergent) thinking, on the other hand, leads in different, and by no means straightforward, ways via erratic chains of association to different approaches to a solution, which can be judged as better or worse (Fig. 3b). This way of thinking has a longer tenure in developmental history; it is echoed in the structure of the human brain. In reality, however, the two ways of thinking can never be completely separated.

2 Creativity Methods

2.1 General Information

Significance: A "method" is a procedure that is built up in logically successive steps and is meant to move from a starting point (task) to a goal (solution) as quickly, safely, and economically as possible. Common examples are a fire escape (from the roof to the ground) or a scheme (completely rigid).

Method and creativity: These terms seem to contradict each other at first. However, the apparent contradiction is resolved when the "ladder" is transformed into a "shelf": The shelf (method) defines the framework (procedure), and creativity fills the compartments (steps) with content (see Jorden, 1977).

Requirements for creativity methods: For unfolding creativity, a method must include the following steps (see Quiske et al., 1973):

- *Collect information:* Tap into as many information memories as possible (starting point for solutions).

- *Remove blockades:* Eliminate inhibiting situations, stuck ideas that are not progressing, etc. (create the possibility for creative solutions).
- *Bring together different patterns:* Increase the probability of a creative leap (solution), even though there is never a guarantee.
- *Evaluate solutions:* Evaluate which approach solves or could solve the problem.

Creative problem solving: This often leads to the goal (often by detours) in the following way:

- *recognize* the problem (i.e., get as close as possible to it),
- *abstract* the problem (i.e., get as far away from it as possible),
- *solve* the problem (i.e., approach it specifically from a distance), and
- *check* the solution (i.e., use logical thinking).

The following creativity methods can be used to support the idea generation process. They are divided into intuitively emphasized and logically emphasized methods (see also Sect. 1.2).

2.2 Intuitively Emphasized Methods

2.2.1 Brainstorming

Significance: Brainstorming (Osborn, 1953) is the oldest, simplest, and best-known creativity method. The version that is widely known and used today was already developed in 1939.

Organization:

- The *group* (team) should consist of approximately 5 to 8 (max. 12) members.
- The participants' *experience levels* should be as balanced as possible (in terms of position, training, education, etc.).
- The participants' *fields of expertise* should be as different as possible.
- The *team leader* acts as a neutral coordinator and navigator.

Procedure: The process takes place in three phases:

1) *Information phase: Explain* and discuss the problem; clarify questions.
2) *Green phase:* Express *ideas* freely with the following "rules":

    - *no criticism* is allowed (this is important),
    - *quantity* is more important than quality,
    - unfeasible ideas are also expected,
    - no competitive behavior (trying to outdo others, keeping ideas to oneself, etc.) is allowed,

- write down ideas in a *visible* location (on a blackboard, flipchart, etc.), and
- look for and develop the *positive* aspects of the ideas (this is important).

3) *Red phase:* Includes the *criticism*, evaluation, and selection of ideas.

- Sort ideas according to their belonging together (numbering on the blackboard, cutting flipchart paper, etc.).
- Discuss and evaluate the ideas (free discussion; if necessary, a simple point evaluation can be conducted).
- Remove useless ideas (e.g., unfeasible ones).
- Select ideas that appear suitable to pursue further.

Comparison with the requirements from Sect. 2.1:

a) *Collect information:*

- Information phase ("problem proximity").

b) *Remove blockades:*

- Prohibiting criticism in the green phase provides psychological security and freedom. The *green–red principle* or principle of pushing back against criticism is the most important rule or behavior in all creative actions.
- Expectation of unfeasible ideas gives participants the courage to express and pursue unusual views.
- Ensuring an even level of experience among the participants prevents them from looking to their superior or other employees for direction (e.g., because they are shy or lack confidence).

c) *Bring together different patterns:*

- Continue ideas through association (open notation).
- Search for positive aspects (every idea has them, but we tend to see the negative ones first).
- "Spinning" ideas can generate new viewpoints.

d) *Evaluate solutions:*

- Red phase (only the first step in evaluation and selection; the project process may start from here if necessary).

Practical experience:

- Brainstorming can be carried out or learned quickly and with little effort, provided that it is treated like "serious play" (like children play); "fooling around" is just as inappropriate as being overly serious.
- Useful results can still be achieved even if the organizational requirements (see above) are not met, provided that the brainstorming is skillfully led.

- It often seems as if, after the end of a session (20 to 30 min maximum), hardly any useful ideas have come out. Only later do the seeds of the ideas that were planted lead to good new solutions after further intensive processing by the person in charge, who, for example, combines and associates the ideas (the activity of the subconscious needs time).
- Useful for the leader is the ability to sketch quickly (practice helps build this ability) and ask stimulating questions.

2.2.2   Brainstorming Variants

Significance: The following variants are intended to expand the possibilities of brainstorming.

Problem reformulation: Ideas for solutions are not immediately sought; first, other formulations for the task (e.g., reversal of the viewpoint, question about the core of the problem, specification) are pursued because the task itself often indicates or restricts the direction of the solution.

Example (Quiske et al., 1973): A hungry monkey sits in a cage with a climbing tree in it; outside the cage lies an unreachable banana. For the monkey, the problem is "How do I get the banana?" A reformulation of this problem could be "How does the banana come to me?" This reformulation directs the brain to other aspects of the problem, which enables new alternative solutions. For example, the monkey might make itself noticed by a guard, who would then hand the banana to the monkey in the cage. Alternatively, the monkey could break a branch off the tree and use it as a tool to reach the banana.

Brainwriting: In this method, each participant writes down three ideas for solutions within five minutes. This means a certain amount of stress for the participants; therefore, brainwriting is sometimes used as an "initial spark" at the beginning of a brainstorming session so that the participants are immediately fully active in the latter part of the session.

The 6–3–5 method (Rohrbach, 1969): This is a written version of brainstorming. It is useful because some people can express themselves better in writing and the ideas are automatically recorded in written form. The procedure for this method is as follows:

- There are **6** participants (composition as in brainstorming).
- Everyone writes down **3** ideas for solutions on a sheet of paper.
- The sheets of paper are then passed around a total of **5** times, with the writing process repeated each time.
- Everyone sees the previous ideas and writes three new ones.

Experiences with the 6–3–5 method: Theoretically, there are $6 \times 3 \times 6 = 108$ ideas across the six sheets of paper at the end of the session. In practice, however, there are fewer ideas because some of them will appear twice and not everyone will always be able to come up with three ideas. Therefore, in addition to the number of participants, the number of ideas can also be changed (e.g., 6–2–5 method or 5–2–4 method).

2.2.3   Image Methods

Significance: Image methods try to make the type and variety of ideas less random than those generated by brainstorming and to intensify the detachment from the problem by analyzing an image. The image is temporarily pushed in front of the problem and thus obscures it (creating distance from the problem and removing blockages); afterwards, an attempt is made to establish relationships between the image and the problem (which brings different patterns together).

Criteria for image selection:

- All participants should *like* the image.
- It must be sufficiently *known* to the participants.
- It must be *far* enough away from the problem area. For example, for a technical problem, an image from nature, and for a social problem, an image from technology is recommended.

Types of image methods:

- *Bisociation method (Koestler, 1966):* Bisociation is the simultaneous viewing of two different (image) ideas. The image should be as close as possible to being the opposite of the problem and should be suggested by the participants according to their own imagination.
- *Analogy method:* The image should, in principle, be related to the problem but come from a distant subject area.

Examples:

1) *Problem:* New umbrella.
2) *Bisociation image:* Setting sun.
3) *Analogy image:* Mushroom.

Recommended procedure for selecting images:

- Participants suggest images (by acclamation).
- The team leader visibly writes down the names of the images.
- The team leader eliminates images that do not comply with the rules (with a short explanation).
- Each participant writes down which image he likes best.
- The team leader can indicate the preferred images (by open vote or using a tally sheet).
- The image with the most votes is chosen; if necessary, a run-off vote is held.

Procedure for image methods:

1) *Explain the problem:* Discuss the problem and clarify questions.
2) *Select the image:* For this procedure, see above.
3) *Describe the image together:*

   - Formulate statements in simple sentences.
   - Write down statements in a visible location (e.g., on a blackboard).
   - No criticism is allowed.

4) *Establish links* between the image statements and the problem:

   - No criticism is allowed.
   - Write down ideas in a visible location.
   - Look for positive aspects and develop the ideas further.

5) Sort, *evaluate*, and *select* ideas (red phase).

Example of a bisociation session (Quiske et al., 1973):

1) *Problem:* Motivating employees to participate in further education courses.
2) *Selected image:* Chick in an egg.
3) *Image analysis* (selection)*:*

   - The chick becomes too large for the shell, and it breaks the shell open.
   - With effort, it leaves the shell and enters a new world.
   - It must stand on its own feet and assert itself.
   - It begins to grow and develop.

4) *Advertising formulations for further education:*

   - Through further education, one can leave one's narrow area of knowledge and advance into new areas.
   - In the course of further education, one takes risks, but these risks provide a chance for completely new professional opportunities.
   - Without further education, one is constricted, dependent, and narrowly viable. One lives almost in a mental straitjacket, like the chick living inside an egg.

2.2.4 Synectics

Significance: Synectics (Gordon, 1961) is a special, sophisticated analogy method. It requires the complete removal of the original problem until it is forgotten. Essential here are techniques related to meditation, such as putting oneself personally in a different situation and the formation of paradoxical word combinations ("every truth is a paradox," i.e., it consists of two opposite poles, see Sect. 1.2). This method requires, even more than image methods, a trained team and an experienced team leader. Training is necessary for the effective execution of this method.

Procedure: The method is carried out in a total of ten steps:

1) Explain the problem and clarify questions. Example: Improve umbrella technically.
2) Search for problem reformulations (see Sect. 2.2.2). Examples: Having hands free, no wet feet, being extremely small (thimble size).
3) Short brainstorming as a "lightning rod": Spontaneously appearing "patent solutions" may block free participation in the further process.
4) Select a problem formulation. Example: Being extremely small.
5) Look for direct analogies to the problem that are related but from a different subject area. Examples: Mushroom, leaf canopy.
6) Select an analogy. It should appeal to all and must be sufficiently familiar. Example: Mushroom.
7) Ask participants to make personal analogies to the chosen analogy: "How do I feel as…?" Examples: Fragile, poisonous, eaten.
8) Form symbolic analogies to the personal analogies; these are paradoxes formed by an adjective and a noun (acting like a book title). Example: Fragile intactness.
9) Seek fantastical and symbolic analogies; any ideas are welcomed (this step can be omitted). Examples: Foam, butterfly wings.
10) Search for solutions by connecting the analogies (especially the symbolic ones) with the problem. This should occur spontaneously and without criticism. If necessary, go through the analogies one by one. Example: An umbrella unfolds like butterfly wings from within the cocoon; although, in this case, the process is to be designed reversible.

Remarks:

- If the results of Step 10 are not satisfactory, go back to Step 6 (choose a new analogy) or Step 4 (choose a new problem formulation).
- For simplicity, Steps 1 to 6 and Step 10 can be followed (i.e., Steps 7 to 9 can be omitted); this is also known in practice as synectics.
- The method can produce highly creative solutions, but it cannot guarantee success.

2.3 Logically Emphasized Methods

Significance: In this section, some more thought-emphasizing methods will be considered; of course, intuition also has its place here. These methods do not necessarily require a team in order to be used.

2.3.1 Analysis of Formulas

Principle: Equations that describe the physical (or other) effect or working principle are investigated. From the consideration of the influencing variables, new (partial) working principles or design elements can be derived.

Example (Pahl et al., 2007): Possibilities for increasing the loosening torque T_L of a bolt or nut are as follows:

$$T_L = F_S \cdot \left[\frac{D_H}{2} \cdot \mu_H + \frac{d_P}{2} \cdot \tan\left(\frac{\rho_F}{\cos(\beta/2)} - \varphi \right) \right] \tag{1}$$

- increase the screw force F_S (by using a high-strength screw),
- increase the head support diameter D_H (by using a screw with a wide head),
- increase the head friction coefficient μ_H (by gluing, material selection, toothing),
- increase the pitch diameter d_P (by using a thick screw),
- increase the flank friction ρ_F (by gluing, dry surfaces),
- decrease $\cos(\beta/2)$ or increase β (by using a screw with a sharp thread),
- decrease the pitch angle φ (by using a screw with a fine thread).

2.3.2 Morphological Box

Significance: Morphology (Zwicky, 1971) is the study of shape or structure. The principle and its technical application are shown in Table 2. The morphological box is usually used only as a two-dimensional morphological table, in which the structure parameters and as many expressions as possible per parameter are listed.

Example: An example of a morphological box is shown in Table 3.

Procedure: The procedure consists of six steps:

1) Explain the *problem*.
2) Divide the problem into *parameters* (e.g., divide each function into subfunctions).

Table 2 Principle and application of morphology

General principle	Exemplary application
A problem is examined to determine its structure, that is, its parameters.	A function is examined for its subfunctions.
Each parameter is displayed in all its expressions if possible.	For each subfunction, all corresponding working principles are listed if possible.
Solution structures result from (all) possible combinations of expressions.	Principle solutions result from possible (compatible) combinations of working principles.

Table 3 Example of a morphological box for a watch. For the functions (top row), appropriate working principles are listed and can be combined to form compatible solutions

Introduce energy	Store energy	Convert energy	Transmit energy	Control energy	Display time
Manual force	Spring	Spring motor	Chain	Pendulum	Hand
Weight force	Weight	Weight motor	Gears	Balance wheel	Folding numbers
Kinetic energy	Compressed air tank	Pneumatic motor	Fluidic circuit	Tuning fork	LEDs
Air pressure fluctuation	Battery	Electric motor	Electronic circuit	Oscillating crystal	Liquid crystal
Electrical network	Sand container	Electromagnet	Light beam	Electronic circuit	Bell signal
Battery	Without	Electronic circuit	Glass tube	Fluidic circuit	Beep

3)   Search for *expressions* for each parameter (e.g., working principles for each function):

- without criticism or valuation,
- as many as possible,
- appropriately categorized according to functional groups (e.g., mechanical, electrical, etc.)

4)   Search for *combinations* of compatible expressions (working principles). This results in approaches for principle solutions.

5)   *Evaluate* combinations (principle solutions).

6)   *Select* one or more combinations for further processing.

Application: The morphological method can be applied to many types of problems that allow for subdivision. The application of this method to functions and working principles shown in Table 2 is typical; it shows similarities with design methodology.

Advantages:

- The method is easy to understand.
- It forces participants to penetrate the problem systematically.
- The method results in a large variety of solutions.
- The green–red principle is followed.
- The method can also be applied by the individual problem solver (designer).

Disadvantages:

- The method is suitable only for cases with relatively few parameters (otherwise, it would be too complex and confusing).

- It always requires an existing or at least imaginable structure. This, however, results in a risk of prefixing.
- The type and variety of solutions are strongly dependent on the parameter distribution. There are any number of different parameter distributions. Therefore, the claim that *all* solutions to problems can be found with this method is not tenable.
- The method does not help when combining several functions into one function carrier ("integral design").

2.3.3 Targeted Feature Variation

Significance: Targeted feature variation (Rodenacker, 1991) is a simple but effective method for systematically developing completely new approaches to solutions starting from an existing design or principle solution. The features under consideration (according to Jorden) are listed in Table 4.

By starting from an overall easier manageability with a focus on a certain part or aspect, a multitude of solution variants can be created in this way. The method is particularly suitable for design activities (from sketching to detailing) if a solution already exists as a principle sketch or drawing but is not satisfactory.

Example: An example of the application of targeted feature variation is illustrated in Fig. 4. Starting with a classical helical spring made of spring steel, first, the *size* is varied by changing the outer diameter of the spring (left image), the wire diameter (center image), or the spring length (right image). A variation in *number* leads to several springs being connected in parallel. The feature *position* can be varied by the position of the force application. In all three cases shown, the force acts as a tensile force. The spring in the left image is loaded in compression, the spring in the center image is loaded in tension, and the two springs in the right image are again loaded in compression. A variation in *shape* results in a conical compression spring (left image), a torsion spring (center image), or a beam spring (right image). As alternative spring *material*, spring bronze (left image) or rubber (right image) are possible. The *surface* can be shot-blasted, polished, painted, or even chrome-plated. Finally, to vary the *connection*, a volume fit in the form of a gas pressure spring (left image) or a form fit via a cylindrical guide (right image) can be used.

Table 4 Targeted feature variation (according to Jorden)

Feature	Variants	Impact
Size Number Position Shape	Geometric	From top to bottom, mostly increasing in deviation from the given design solution
Material Surface Connection	Other	

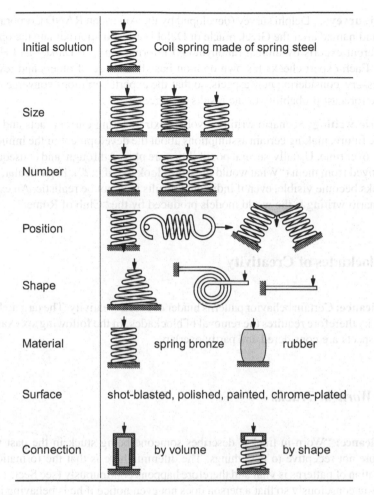

Initial solution	Coil spring made of spring steel
Size	
Number	
Position	
Shape	
Material	spring bronze rubber
Surface	shot-blasted, polished, painted, chrome-plated
Connection	by volume by shape

Fig. 4 Targeted feature variation in a coil spring (according to Jorden)

Practical experience: The targeted search in different, predetermined directions leads relatively quickly away from prefixed ideas. By adjusting and combining ideas further, new solutions can be found with little effort—possibly after several attempts. The method can be applied both in a group and by individuals.

2.3.4 Futurological Methods

Significance: Futurological methods help derive future prospects from existing facts and views. This can make future problems, whose solutions can or must be prepared today, visible.

Delphi survey: A Delphi survey (developed by the American RAND Corporation in 1963 and named after the Greek oracle of Delphi) aims to approximate the opinions of different experts by mutual (anonymous) information (see Dalkey and Helmer, 1963). Each expert checks his own opinion based on those of others and revises it if necessary, considering new aspects, so that the experts' opinions converge overall and the forecast probability of the results is increased.

Scenario writing: Scenario writing involves extrapolating current facts and trends into the future, making certain assumptions about the development of the influencing factors over time. Usually, several possibilities are played through, and consequences are derived from them ("What would the world look like if…?"). In particular, trends and risks become visible, even if individual results may not be realistic. An example of scenario writing is the world models produced by the "Club of Rome."

3 Blockades of Creativity

Significance: Certain behavior patterns hinder or block creativity. The targeted use of creativity therefore requires the removal of blockades. In the following six examples, such aspects are considered and partly overlap.

3.1 Worn-in Tracks

Significance: "Worn-in tracks" describes someone being stuck in the past who is therefore not receptive to new things. The dilemma here is that the formation and application of patterns is vital and therefore happens continuously (see Sect. 1.2) and always unconsciously so that a person does not even notice if he is behaving in such a way.

Category formation: Human consciousness divides patterns into categories. These enable or facilitate the adaptation of behavior to the demands of life; however, they also determine the mobility and availability of patterns:

- Categories enable the rapid *identification* and *classification* of perceptions into existing categories. The risk here lies in premature (prejudiced) judgment. For example, if one notices a man staggering, one might pass the judgement that the man is drunk. However, it could be that he is feeling faint and needs help.
- Human consciousness always tries to move forward with the *smallest possible number* of categories and to maintain them. Rearranging and creating new categories requires mental effort and causes anxiety because a part of our worldview that we are used to must be given up. "Comfortable" retention means immobility, "calcification," conflict aversion, the renunciation of creativity, and, finally, the

further development of one's personality. An extreme example (or caricature) is that someone medicates everything he can see with iodine and what he cannot see with laxatives.

- Categories usually contain a correct foundation but are often not very rational; in addition, they are usually emotionally colored. Even an originally meaningful classification becomes inappropriate after some time due to a lack of adjustment or is inadmissibly generalized. This leads to prejudices and objectively unjustified views, which are the enemy of personality development. The surest way to keep categories secure is to follow the saying, "Exceptions prove the rule."

Approaches to escape from worn-in tracks: The central task is to become aware of this situation (the goal of all personality development). This is done through the following:

- Questioning preconditions:

  - What is truly important?
  - What is the real problem?
  - Which conditions are actually given, which are not?

- Gathering many suggestions and consciously processing them.
- Searching for new aspects and not suppressing them as well as checking categories and, if necessary, creating new ones.
- Training one's mental agility (see Sect. 4), that is, playing with words, concepts, and ideas (keeping away from clichés).

3.2 Obedience to Authority

Significance: A person obedient to authority is someone who blindly, that is, without questioning, accepts everything that a "higher" authority (e.g., parents, a work superior, an expert, the state, a political party, a church) declares to be right. Even real authorities can be mistaken; everyone always sees a problem from his own point of view.

Example: Poincaré "proved" logically that, due to the curvature of the earth, it would be impossible to radio around the Atlantic Ocean. However, Marconi tried still it and succeeded (always check the preconditions).

Authority: There are two terms to distinguish here:

- *authority by virtue of office* (holding authority = power),
- *personal authority* (being an authority); this includes

  - professional competence and character integrity (prerequisite) and
  - recognition by the others (realization).

Personal authority is seen in quite a positive light. It cannot be achieved willingly (if someone tries, he will achieve exactly the opposite). Children, for example, need it as a model. A rejection of authority means only not accepting anything outright without *reflection* and consciously making up one's *own* mind.

Approaches to avoiding obedience to authority:

* Question one's own opinion:

  - Where does it come from?
  - Is it proven?
  - Is it one-sided or does it need to be supplemented by other views?

* Turn generally accepted views upside down and draw conclusions from them ("What would the world look like if…").
* Not only search logically but also listen to one's own "inner voice" (intuition).

3.3   Conformity

Significance: Conformity means being oriented around the values set by one's environment (a group, the general public). This results in adaptations to become more ordinary and thus in a restriction of one's own field of vision. Creativity, on the other hand, always means countering what is "usual" with something new. Conformity is thus related to obedience to authority and, in a broader sense, to "worn-in tracks."

Extreme examples:

* The Reichstag vote on "total war."
* "Three-line comparison" (Asch, 1951): This is a test situation in which an (uninformed) test person sits opposite an (initiated) group. The test person is shown three lines of clearly different lengths as well as a reference line, the length of which clearly resembles one of the three lines. Each member of the group, ending with the test person, states which line the reference resembles. In approximately two-thirds of the cases, the group will express a unanimous wrong opinion. Past experiments show that approximately one-third of the test persons will be "tipped over" if more than half of the other group members select the wrong line and will agree with the visibly wrong group opinion. Another third, on the other hand, will not be persuaded. As soon as even a single group member selected the same line as the test person, the number of "tip-overs" is almost zero.

The examples show that consciousness can be manipulated, which is comparable to the alignment of magnetic dipoles in soft iron when a magnetic field is applied.

Approaches to avoiding conformity:

- This situation is similar to obedience to authority (see above) and can thus be combatted similarly.
- Awareness of the processes is crucial.

3.4   Emotional Insecurity

Significance: New ideas are uncomfortable. *Courage* (from the individual person) and *mental space* (from the group) are needed to express them. These needs influence each other. The individual can produce ideas toward a goal (problem solution) only as long as he is emotionally free enough, that is, as long as he does not have to use his energy for personal actions or reactions (see the following example).

Example problem meeting (Quiske et al., 1973): Ideally, according to Fig. 5a, all activities are factually directed toward the goal. However, each participant views the problem only from his own perspective. An approach is possible only if everyone is prepared to change their "point of view" at least occasionally, to go "deeper" into the problem, or to put himself "above the issue."

In an unideal case, according to Fig. 5b, a personal "attack" (e.g., criticism) by Person A leads to Person B feeling "hit," unconsciously going into a "defensive position," building up a "protective shield" that at least partially obscures his view of the problem, feeling irritated (mostly unconsciously), and eventually "shooting down" another person who was previously uninvolved in the process, for example, Person C. After some time, everyone becomes aggressive and takes things personally; the problem is then pushed out of focus.

Further examples:

- Plenary debate in a parliament.
- Creativity studies of school classes have shown that a class with a strict teacher is *less* creative than other classes but also shows less conscious fear of school.

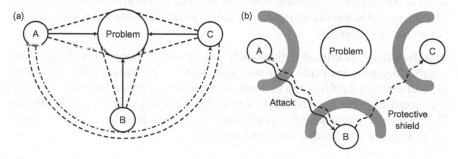

Fig. 5 Problem meeting: **a** problem-related, **b** with personal attacks. Based on Quiske et al. (1973)

The repression mechanisms that suppress fear apparently occupy a considerable part of the "working memory," like an "operating system," and block essential "files" with their "patterns" (especially unconventional ones, which are important for creativity).

Approaches to combatting emotional insecurity:

- Applying creativity methods with the green–red principle
- Creating a creative environment by practicing the green–red principle in everyday life (see Sect. 3.6)

3.5 Egocentricity

Significance: Egocentricity means always centering oneself. It represents an exaggeration of egoism (*self*-centeredness), which is, to a certain extent, vital (self-esteem and self-love are basic prerequisites for personality development). However, in principle, the egocentric sees only the ego in the mirror and not the problem. Egocentric behavior also includes status or role thinking ("A boss cannot express something like that"). This problem is similar to those of "conformity" and "emotional insecurity."

Approach to combatting egocentricity: This is difficult (question of personality development; it is difficult to make the egocentric person aware of this).

3.6 Critical Basic Attitude

Significance: A critical basic attitude should be understood as the *premature* condemnation (*pre*judgment) of new ideas due to negative aspects. Everything that exists has positive and negative aspects (it is a question of viewpoints and is not absolute); we are often inclined to see and point out the negative ones first. This often means the death of the idea and, at the same time, offence to the originator. This does not mean that one should be uncritical; rather, one should always consider the positive aspects before passing judgment.

Manifestations: A critical basic attitude can be seen, among other things, when negative demands are made ("We no longer want…" or "Down with…") without the will to return the favor, not to mention any advance performance (this is always necessary). Even having "too much" knowledge can lead to a critical basic attitude: The "specialist" immediately sees only concerns and risks in new ideas (this is similar to the problems stemming from "worn-in tracks").

Approaches to avoiding critical basic attitude:

- Always formulate demands or wishes *positively.*
- Consciously adopt the *green–red principle.*
- *Listen* to other people (this requires conscious practice; usually, we listen only until the other person says something we disapprove of).
- In the case of diverging opinions (e.g., expert meetings), look for *commonalities*; if necessary, look at the issue from a "higher perspective" (see Fig. 5).

4 Promotion of Creative Behavior

4.1 Characteristics of Creative People

Significance: Psychological studies show that there are a number of qualities that can be found within creative personalities but none that can be considered an absolute prerequisite. In this context, Guilford (1967) defined six "cognitive abilities" of a creative person ("cognitive" means related to the area of perception, thinking, and memory). Other psychologists use both similar and different terms or classifications. Such abilities can be practiced in a similar way to training and keeping one's body flexible through gymnastics or sports.

1. **Sensitivity:** Ability to grasp problems; open-mindedness toward the environment.
2. **Thought awareness:** Ability to activate memory material (patterns) in:

    - *words* (word familiarity);
    - *associations:* making connections of content;
    - *expressions:* formulating mental content;
    - *ideas:* finding, for example, meaningful terms for certain conditions.

3. **Mobility (flexibility):** Ability to leave familiar paths, shift existing thought patterns, and do so:

    - *spontaneously:* changing between different points of view; and
    - *adaptively:* adjusting to the respective requirements of the situation without being too attached to the previous one.

4. **Originality:** Ability to see things differently, make unusual associations, find appropriate terms.
5. **Redefinition:** Ability to abandon habitual views of familiar objects and define and use them differently.
6. **Elaboration:** Ability to create a structure or plan from given information and expand the details included in it.

4.2   Guiding Rules for Creative Behavior

Significance: The following are some guidelines for creative behavior, the application of which can have a positive effect on all areas of work. These rules are not considered to be complete, nor does their overly strict application make sense. Rather, they are intended to provide useful tips for a creative approach.

1 Process
Acquire and convey knowledge of the creative process.

2 Listen
Listen to other people as well as one's own intuition without making immediate judgments.

3 Positive
Always consider the positive aspects of new ideas first.

4 Thinking patterns
Do not use or enforce rigid thinking patterns.

5 Playfulness
Playfully deal with objects as well as with thoughts.

6 Intimidation
Do not feel intimidated by the achievements of others.

Anxiety is a blocker. Do not forget how long it takes to create a masterpiece.

7 Stimulation

Encourage others to be creative (through a creative atmosphere, tricky questions, the acceptance of ideas).

Sparking creativity in others is a creative skill that is especially important for leaders.

8 Personalities

Promote and encourage creative personalities to avoid the unwillingness of others.

Nonconformative thinking does not mean rebelling against conventions (i.e., the rules of the game that facilitate daily interaction). Strong personalities in particular should consciously place themselves within the general framework; otherwise, too much energy is lost to useless action.

9 Phases

Alternate between active and passive phases.

10 Realization

Note down ideas and realize them or make sure they are realized.

11 Knowledge

Acquire knowledge and skills in many areas (other than one's specialized field).

12 Group work

Acquire knowledge of human behavior in groups, encourage and practice group work.

References

Asch, S. E. (1951). Effects of group pressure upon the modification and distortion of judgment. In H. Guetzkow (Ed.), *Groups, Leadership and Men* (pp. 177–190). Pittsburgh: Carnegie Press.

Dalkey, N., & Helmer, O. (1963). An experimental application of the Delphi method to the use of experts. *Management Science, 9*(3), 458–467.

Gordon, W. J. J. (1961). *Synectics: The Development of Creative Capacity.* New York: Harper.

Guilford, J. P. (1967). *The Nature of Human Intelligence.* New York: McGraw-Hill.

Jorden, W. (1977). Ist Kreativität erlernbar? *Paderborner Studien, 3/4*(1/2), 44–51.

Jorden, W. (1988). Zur Ausbildung des Menschen im Studiengang Konstruktionstechnik. In V. Hubka (Ed.), *Proceedings of the International Conference on Engineering Design* (Vol. 1, pp. 459–465). Zurich: Heurista.

Koestler, A. (1966). *The Act of Creation.* London: Pan.

Osborn, A. F. (1953). *Applied Imagination: Principles and Procedures of Creative Thinking.* New York: Scribner.

Pahl, G., Beitz, W., Feldhusen, J., & Grote, K.-H. (2007). *Engineering Design. A Systematic Approach* (3rd ed.). London: Springer.

Poincaré, H. (1913). *The Foundations of Science.* New York: Science Press.

Quiske, F. H., Skirl, S. J., & Spiess, G. (1973). *Denklabor Team – Konzept für kreative Problemlösungen in Forschung, Verwaltung und Industrie.* Stuttgart: Deutsche Verlagsanstalt.

Rodenacker, W. G. (1991). *Methodisches Konstruieren. Grundlagen, Methodik, praktische Beispiele* (4th ed.). Berlin: Springer.

Rohrbach, B. (1969). Kreativ nach Regeln – Methode 635, eine neue Technik zum Lösen von Problemen. *Absatzwirtschaft, 12*(19), 73–75.

Taylor, I. A. (1959). The nature of the creative process. In P. Smith (Ed.), *Creativity: An Examination of the Creative Process* (pp. 51–82). New York: Hastlings House.

Zwicky, F. (1971). *Entdecken, Erfinden, Forschen im Morphologischen Weltbild.* Munich: Droemer Knaur.

Evaluation and Selection Methods in a Product Development Project

1 Overview

Significance: Evaluation and selection activities always occur where there is more than one idea or solution. Together with creation activities, they form the core of the problem-solving process. The methodological basis for this is provided by "VDI 2225 Part 3" (1998).

Evaluation methods: Summary of the information required or available for evaluation using a uniform scheme; condensation of the information into (usually) one or two value figures or a value judgment. A distinction is made among

- verbal methods,
- point value methods, and
- parameter methods.

Verbal methods: These are the simplest type of evaluation (see Table 1 for an example). Criteria or questions are listed (e.g., by market or producibility) and then assessed or answered verbally; the overall assessment is created intuitively by weighting the individual assessments.

Point value methods: In these methods, the evaluation criteria are first assessed numerically using a points scale, and then the values are added up (with or without weighting); see Sect. 2.

Parameter methods: Here, the evaluation is made by assessing individual economic parameters (see "Systematische Produktplanung" 1983).

Application of the methods: Various methods can be assigned to the different phases of product development, as shown in Table 2.

J. Schlattmann and A. Seibel, *Structure and Organization of Product Development Projects*, https://doi.org/10.1007/978-3-030-81046-7_7

Table 1 Example of a verbal assessment

Criteria	Assessment
Market	Large, growing
Manufacturing	New machine necessary
…	…
Overall assessment: promising	

Table 2 Application of evaluation methods

Phases	Methods		
	Verbal methods	Point value methods	Parameter methods
Planning	×	×	
Designing		×	
Elaborating		×	×

2 Point Value Methods

2.1 Simple Point Evaluation

Basis: The evaluation criteria are based on the requirements list; they usually contain the following:

- fixed demands (these must be met),
- limited demands (these must be met within the limits), and
- wishes (these should be fulfilled, depending on the effort required).

Procedure for point evaluation:

1) Establish *criteria* that

 - cover all essential characteristics (requirements list),
 - are roughly balanced (no unimportant ones),
 - are independent of each other (i.e., it should be possible to assign any property of the product under assessment to exactly one of the criteria),
 - are positively formulated (e.g., "low costs" instead of just "costs"),
 - total approximately 6 to 10 (there should not be too many).

 An example of a criteria list is given in Table 5.

2) Determine the *weight W* of the criteria. This step can also be omitted and all criteria be equally weighted (which is often sufficient, as the weighting is noticeable only if highly weighted criteria have very high or very low degrees of fulfilment; see Point 4). The following methods may be used:

Table 3 Point scale for evaluation. Based on "VDI 2225 Part 3" (1998)

Points	Definition	Meaning
$F = 4$	Excellent	F_{max}, rare top quality (ideal)
$F = 3$	Good (+ or A)	Normal range
$F = 2$	Average (○ or B)	
$F = 1$	Bad (− or C)	
$F = 0$	Unbearable	Blocks the solution (fails)

Table 4 Meaning of rating (reference)

Rating R	Meaning
>0.8	Very good
0.8 to 0.7	Good
0.7 to 0.6	Useful
<0.6	Insufficient

- Intuitively estimate which criteria should be weighed more heavily and assign them factors (e.g., $W = 2$) or
- Estimate the weights using a weighting matrix (see Sect. 2.3).

3) Specify a *points scale*: The scale seen in Table 3 has proven to be particularly useful. Other scales are also possible (e.g., $F = 1 \ldots 10$, $F = 1 \ldots 100$, school grades).

4) Determine the *level of fulfilment F*: For each criterion, estimate how well the individual solutions meet it. It is useful to go through the criteria individually and compare the solutions with each other.

5) Calculate the *total points* $P = \sum W \cdot F$: The solution with the highest sum P is relatively the best; however, this does not say anything about its absolute suitability (it may still be a bad solution).

6) Calculate the *rating R*: The achieved score P is related to the maximum possible score $P_{max} = n \cdot F_{max}$ (n = number of criteria): $R = \frac{P}{P_{max}}$

7) Assess the *rating:* The numerical values R can be assigned approximately (but caution is warranted) to the meanings according to Table 4; however, the terms are to be understood only as reference values.

Example: Table 5 shows a (not exemplary) evaluation of three designs for modular gear units.

2.2 Economic Rating

Significance: The economic rating is one aspect of value determination (see Sect. 2.1). It is considered separately because it can be handled in different ways:

Table 5 Exemplary point evaluation of gearbox designs ($F_{max} = 4$, $P_{max} = 40$)

Criteria	W	Design A		Design B		Design C	
		F	$W \cdot F$	F	$W \cdot F$	F	$W \cdot F$
Small number of different cast parts	2	1	2	2	4	3	6
Small number of different gear wheels	1.5	3	4.5	3	4.5	1	1.5
Space-saving design	1.5	4	6	3	4.5	2	3
Low weight of gearbox	1	3	3	2.5	2.5	1	1
Transmission ratios per housing	1	2	2	2	2	3	3
Simple stock keeping	1	1	1	3	3	3	3
Favorable stock production	2	1	2	2.5	5	3	6
Sum			20.5		25.5		23
Rating R			0.51		0.63		0.58

a)   Economic aspects (costs) can be included in the *criteria list*. Then, the weighting or the number of criteria affect the extent to which the evaluation and selection of solutions is determined by function-related (i.e., technical) or cost-related (i.e., economic) aspects.

b)   Economic criteria can be treated *separately* (see below). According to "VDI 2225 Part 3" (1998), the technical R_t and economic rating R_e are determined individually and then compared. This allows, for example, the tendency toward "luxury" or "cheap" products to be controlled.

c)   In the *cost–benefit analysis* (see Sect. 2.3), the criteria can be divided into "technical" and "economic" ones and can also be tracked separately.

Economic rating according to "VDI 2225 Part 3" (1998): This rating includes only the manufacturing costs and not the operating costs; the latter must be taken into account in the technical criteria. The difficulty here lies in finding a reference to the "ideal" solution (solution with F_{max}).

Procedure for determination:

1)   Determine the *lowest comparable market price* $P_{m,min}$ (from market research or similar). The problem here lies in the comparability.

2)   Determine the *cost factor* β from the operating account: $\beta = \frac{\text{market price } P_m}{\text{manufacturing costs } M}$ (rough reference value: $\beta \approx 2$)

3)   Determine the *"permissible" manufacturing costs* for the new product: $M_{per} = \frac{P_{m,min}}{\beta}$

4)   Determine the *"ideal" manufacturing costs* (experience value): $M_{ideal} = 0.7 \cdot M_{per}$

5)   Determine the *economic rating* of the new product: $R_e = \frac{M_{ideal}}{M_{design}}$

Fig. 1 Ratings diagram.
Based on "VDI 2225 Part 3"
(1998)

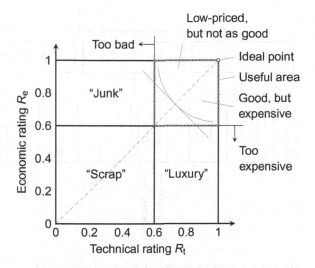

Comparison of the ratings: If the technical and economic ratings are determined separately, they can be graphically compared in a ratings diagram (see Fig. 1). Balanced designs will appear approximately on the diagonal. Solutions that are approximately equal to each other lie on circles around the ideal point or, according to another idea, on straight lines perpendicular to the diagonal.

2.3 Cost–Benefit Analysis

Significance: A cost–benefit analysis (Zangemeister, 1970) is a *refined* point evaluation. The following procedure essentially captures the differences from the simple point evaluation described in Sect. 2.1.

Procedure:

1) *Criteria plan:* The evaluation criteria are subdivided hierarchically; that is, first, a few main criteria are established, and then, they are further subdivided (see Fig. 2 for an example). This makes it easier to keep track of the criteria and keep them independent.

2) *Weighting of the criteria:* The criteria are weighted in such a way that the sum of the weightings at each level is 100% (or 1) (for an example, see Fig. 3). This results in the following:

   - The *importance* of the main criteria is visible (e.g., emphasis on low costs).
   - The criteria can be directly *compared* with each other at each level, especially at the lowest one.

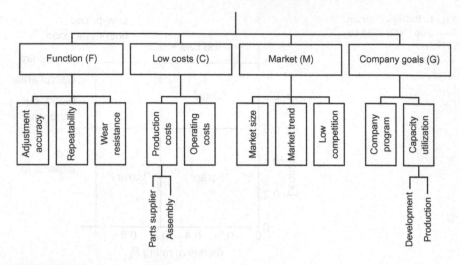

Fig. 2 Example of a hierarchically structured criteria plan

Fig. 3 Example of a weighting for Fig. 2

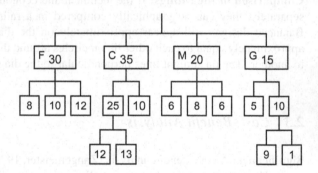

The weighting can be based on free estimation or be done by using a *weighting matrix* (see Table 6):

a)   Within a group of criteria, each criterion is compared with all of others individually.

Table 6 Example of a weighting matrix

Criteria		F	C	M	G	$\sum$	G (preliminary) (%)	G (final) (%)
Function (F)	F	(F)	C	F	F	3	30	30
Low costs (C)	C	C	(C)	C	C	4	40	35
Market (M)	M	F	C	(M)	M	2	20	20
Company goals (G)	G	F	C	M	(G)	1	10	15
Total						10	100	100

Table 7 Example of an evaluation table (excerpt)

Individual criteria	Degrees of fulfilment (evaluation levels)				
	$F = 4$	$F = 3$	$F = 2$	$F = 1$	$F = 0$
Adjustment accuracy	$\leq 2\,\mu m$	$>2 \dots 4\,\mu m$	$>4 \dots 7\,\mu m$	$>7 \dots 10\,\mu m$	$>10\,\mu m$
Repeatability	$\leq 1\,\mu m$	$>1 \dots 2\,\mu m$	$>2 \dots 3\,\mu m$	$>3 \dots 5\,\mu m$	$>5\,\mu m$
Wear	None	Low	Average	Slightly heavy	Heaviest

b) In each case, the criterion that is estimated to be higher is entered in the matrix field (and the criterion itself can be found on the diagonal).

c) The number of mentions of each criterion in its own row is a measure of its importance.

d) The numbers are proportionally extrapolated so that their total sum is 100% (provisional weighting).

e) The final weightings are freely adjusted from the provisional ones so that the total is back to 100%. The reason for this is that the comparison under Point a does not yet indicate the numerical ratio.

When weighting criteria at lower hierarchical levels, only those criteria that belong to the same superordinate criterion are compared with each other (e.g., the three subcriteria for "function" in Fig. 2). The sum of the weightings for all subcriteria is equal to the weighting of the superior level.

Only the criteria of the lowest, that is, most detailed level, are used for the subsequent evaluation.

3) *Points scale:* Same as in Sect. 2.1: $F = 0 \dots 4$ (or $F = 0 \dots 10$; $F = 0 \dots 100$) with intermediate values, if necessary.

4) *Levels of fulfillment:* Before the evaluation, it is advisable to determine for each criterion which characteristic of the product property is to be evaluated with how many points (see Table 7 for an example). This provides a basis for comparison among several evaluators or for later evaluations.

The following steps are schematic and correspond to the procedure described in Sect. 2.1.

5) *Benefit value* = points total: $P = \sum W \cdot F$
6) *Rating* (relative benefit value): $R = \frac{P}{P_{max}}$
7) *Assessment:* For indicative values, see Table 4.

Advantages of the cost–benefit analysis:

• It forces the systematic penetration of the evaluation.
• The effort required is relatively low, especially when using forms or corresponding software (e.g., Excel).

- The necessary information can usually be obtained in-house.
- The weighting of the main criteria makes it possible to control the direction of product development (e.g., cheap or luxury product).
- The results are transparent, that is, one can see why one product performs better or worse than others.

Disadvantages of the cost–benefit analysis:

- The criteria, weightings, and levels of fulfilment are always subjective.
- The independence of the criteria is not always guaranteed.
- The schematization can falsely appear to be exact.

Therefore, one should not look only at the numerical result of the rating. Poorly evaluated properties, that is, those with low levels of fulfillment, indicate starting points for product improvement.

If necessary, potential higher fulfillment levels can be noted in the evaluation procedure (e.g., in brackets).

Example: Fig. 4 shows the application of a cost–benefit analysis in the evaluation of two propulsion concepts for a city vehicle.

CRITERIA PLAN					SOLUTION EVALUATION			
① Main criteria	② Weighting W	① Individual criteria	Abbrev.	① Weighting W	Solution 1: Combustion propulsion		Solution 2: Electrical propulsion	
					F	F·W	F	F·W
Power	20	Velocity		7	4	28	1	7
P		Acceleration		6	2.5	15	3	18
		Range		7	4	28	1	7
Profitability	30	Low purchase price		9	3	27	2	18
C		Low energy costs and taxes		12	2	24	2.5	30
		Low maintenance costs		9	2	18	2	18
Operation	15	Simple operation		3	2	6	3	9
O		Little force effort		2	3	6	2	4
		Availability		5	3.5	17.5	1.5	7.5
		Reliability		5	3	15	4	20
Environmental	35	Low exhaust emission		15	1	15	3	45
sustainability		Little noise generation		10	1.5	15	4	40
E		Recyclability		10	3	30	2.5	25
Total	100			∑W 100				
		⑤ Points total P (benefit value)			244.5		248.5	
		⑥ $P_{max} = F_{max} \cdot \sum W \Rightarrow$ Rating $R = \frac{P}{P_{max}}$			0.61		0.62	

② Weighting matrices

Criteria \ Criteria	1	2	3	4	5	6	∑	W_prel.	W_final
	P	C	O	E					
1 Power	P	P	C	P½	E		1.5	15	20
2 Profitab.	C	C	C	C	E		3	30	30
3 Operat.	O	O½	C	O	E		1.5	15	15
4 Env. sust.	E	E	E	E	E		4	40	35
5									
6									
Totals							10	100	100

Criteria \ Criteria	1	2	3	4	5	6	∑	W_prel.	W_final
1									
2									
3									
4									
5									
6									
Totals									

③ Points scale and ④ Fulfillment levels	Fulfillment levels (evaluation levels)				
	$F_{max} = 4$	$F = 3$	$F = 2$	$F = 1$	$F = 0$
④ Individual criteria \ ③ General signif.	Excellent (ideal)	Good	Average	Bad	Unacceptable (fails)
Velocity	> 100	80...100	70...80	60...70	< 60
Range	> 200	150...200	100...150	60...100	< 60

Fig. 4 Example of an evaluation of two propulsion concepts for a city vehicle (excerpt)

References

Systematische Produktplanung. (1983). Dusseldorf: VDI.

VDI Guideline 2225 Part 3. (1998). *Design engineering methodics—Engineering design at optimum cost—Valuation of costs.* Berlin: Beuth (in German).

Zangemeister, C. (1970). *Nutzwertanalyse in der Systemtechnik.* Munich: Wittemannsche Buchhandlung.

Leading Employees in a Product Development Project

1 Basics

Definition: Leading means influencing others in such a way that they change their location (physically or mentally) in a desired way (according to Schleip; see Ammelburg, 1993).

Theses on leadership: Leading is more than "instructing." The following statements apply:

- *Leadership* can be learned and must be learned like all other professional qualifications.
- *Mistakes* in leadership can cause far greater damage than mistakes in the calculation and design of products.
- *Leadership training* is usually far too short in usual university education.

Styles of leadership: Typical leadership styles are:

- *Authoritarian leadership* (commanding):

  - Disadvantages: Little self-development and restriction of initiative.
  - Leads to dissatisfaction, aggression, and intrigues.
  - Today, largely not an option.

- *Laissez-faire leadership* (nonintervention):

  - Disadvantages: Divergence of initiatives, lack of coordination and community.
  - Leads to dissatisfaction, confusion.

- *Cooperative leadership* (motivating):

  - The employee is (despite the hierarchy) seen and addressed as an equal and of equal rank (measures are explained; commands are used only as a last measure in case of wrong developments).

J. Schlattmann and A. Seibel, *Structure and Organization of Product Development Projects*, https://doi.org/10.1007/978-3-030-81046-7_8

- This leads to self-development, initiative, harmony, and positive cooperation internally and externally.

Motivation: Human attitudes often dictate that someone will do something only out of self-interest and inner need. Motivation is possible only if the leader has knowledge of the following:

- *human nature* and
- basics of *applied psychology*.

Elementary mistakes in these points—mostly made unconsciously—are the cause of most cases of failure in leadership tasks.

Remark: It is not possible to teach sound knowledge of human nature in a single chapter. Nevertheless, an attempt is made here to give some suggestions that have proven themselves in practice; these are suggestions for individuals' own further training and personality development, but above all, they are guidance in at least recognizing the worst and most frequent leadership mistakes and knowing how to deal with them as well as possible.

2 Structure of Human Nature

2.1 Introduction

Knowledge of human nature in everyday use: It usually happens unconsciously, for example, after

- a first impression (this may be meaningful, but caution is warranted, as it may be distorted by, e.g., current external circumstances or mood),
- sympathy or antipathy (this is an individual reaction, and there is no specific criterion for it),
- momentary actions (these are formed in snapshots, so they have no general validity),
- observing someone's external appearance (e.g., high forehead or long nose).

In this way, knowledge of human nature is not possible because this leads to unfounded, hasty judgments, which are equivalent to prejudices (they are the worst enemy of personality development and are strengthened by the attitude that "exceptions confirm the rule"). Nevertheless, the external characteristics of a person allow another person to apply his knowledge of human nature to their interactions to a certain extent, which can be helpful under certain circumstances.

Models of the human being: The human being is one of the most complicated entities in nature. It cannot be easily analyzed and represented like a product structure. Nevertheless, psychologists and physiologists have developed various "models" of human beings. These models have been confirmed by experience and statistical studies but are, to some extent, controversial.

"Onion model" according to Endres (1954): In this model, humans' external characteristics and various areas of internal essence are like onion skins around the essence (see Fig. 1). The "consistency of the essential characteristics" means that the wholeness of a human being is based on a uniform "blueprint" according to which the external characteristics and internal essence areas are causally connected. If it is possible to determine a human's "type" according to his external characteristics, then, according to this idea, well-founded conclusions can be drawn about his way of thinking, feeling, etc., but with the following limitations (see Ammelburg, 1993):

- There is a difference between a natural predisposition and personal development.
- There are no clear external appearance types, only three different basic types of humans (see below).
- Some branches of modern psychology deny the demonstrable connection between the physical and the psyche, while others confirm it.

Dominance of the layers: The darker the layers in Fig. 1 are, the more completely they belong to the unconscious, that is, the *"deeper"* they are. The deeper lying layer is always *stronger* than the layer surrounding it. For example, arguments (logical thinking) are of no use against love (feeling or experiencing). The unconscious reaches into *all* areas, including the outer ones.

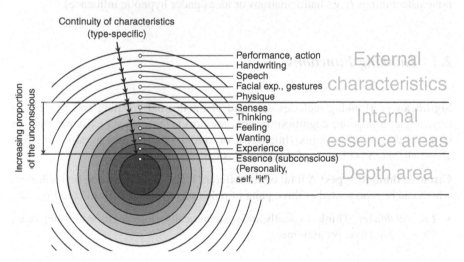

Fig. 1 Onion model of the structure of the human being. Based on Endres (1954), see also Ammelburg (1993)

In the following, the layers of the inner essence areas are considered, as far as they are essential for leadership psychology (see Ammelburg, 1993).

2.2 Sense Area

Significance: The senses include sight, hearing, smell, taste, touch (contact, temperature), and more, such as the sense of balance and position (limb position). Senses are entrance gates to humans' inner existence; the more they are triggered, the stronger the effect. Approximately 80 to 85% of humans' sensory impressions enter the brain through the eye, and approximately 10 to 15% enter through the ear. Therefore, illustrative material is important (e.g., consider the impact of television versus that of radio).

Connections: Sensory functions are never addressed in isolation; other essence areas are always involved (thinking, feeling, wanting, experiencing; e.g., looking at an accident). This may result in sensory illusions (the inner superposition of deeper layers):

> **1 Senses**
> Senses are *deceivable*.

This deception can even go as far as the concrete perception of objectively nonexistent things (e.g., hallucinations or ideas under hypnotic influence).

2.3 Thinking Functions

Significance: Thinking includes any kind of intellectual grasping and processing (e.g., logic, reflection, cognition, reason, memory). Differences exist not only in terms of the *ability* to think (intelligent or stupid—therein lies a valuation) but, above all, in terms of *ways* of thinking (this is *value-free*).

Circle of thought types: A total of six different types of thinking can be identified, which can be represented as three pairs of poles set in a circular form (see Fig. 2):

- *Logical thinker:* Thinks logically and systematically; does not miss any step (e.g., lawyer, detective, programmer).

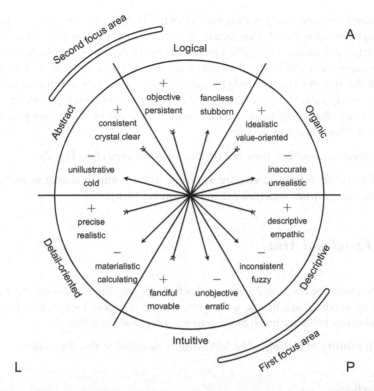

Fig. 2 Circle of thought types (with subjective value judgments) with the three basic types: A, P, and L (see below). Based on Schleip (see Ammelburg, 1993)

- *Organic thinker:* Also thinks logically but directs the view more to the whole as an organically connected complex; may even skip a step.
- *Descriptive thinker:* Must perceive everything with the senses; must be able to use his imagination.
- *Intuitive thinker:* On the opposite pole from the logical thinker and therefore erratic and creative; inventor type [for an evaluation, a logician is often also necessary; e.g., architect (intuitive) + structural engineer (logical)].
- *Abstract thinker:* On the opposite pole from the descriptive thinker; does not need a view; captures problems on their own (e.g., mathematician, philosopher, theologian).
- *Detail-oriented thinker:* On the opposite pole from the organic thinker; analyzes the issue in the smallest possible detail and dissects problems; the complex whole interests him less (e.g., researcher, book reviewer).

While the areas of most thought types in Fig. 2 flow smoothly into each other, there is a kind of border between detail-oriented and intuitive thinking (jumping from the smallest detail to the detail-free whole). This results in a field of mental tension.

Individual focuses: Every person has his individual focus somewhere in the circle of thought shown in Fig. 2. This focus can vary in pronunciation and width. Training and practice can increase its width. Operating toward the opposite pole, however, will always cause more difficulties than operating in one's own focus area. Accordingly, there is the risk of (unconsciously) regarding one's own way of thinking as "right" and negatively judging people who use the opposite way of thinking; both tendencies, however, are objectively wrong (marked with + and – in Fig. 2). When people "talk past each other," this may be why.

Focus areas: Altogether, there are two major focus areas (see Fig. 2):

1) *descriptive–intuitive* area (largest group; more common among women) and
2) *logical–abstract* area (more common among men).

2.4 Emotional Area

Significance: The emotional area includes feelings, sentiments, moods, etc. Feelings are partly unconscious; that is, we are often not clear about their exact cause. They determine our behavior much more than we often want to admit.

Transferability of feelings: The following is essential to this discussion:

> **2 Feelings**
> Feelings are "contagious"; that is, they transfer directly to others.

Approaching other people with a positive, open-minded basic attitude can help transfer this attitude (directly or indirectly) to one's counterpart and therefore lead to meeting with many more positive reactions than doing so with a negative attitude.

Conversely, it is (unfortunately) also true that a negative basic attitude can be detrimental to human interactions:

> **3 Negative basic attitude**
> He who approaches the things in life with a wrong, negative attitude will always find his wrong attitude confirmed in the end.

This process corresponds to a wrongly polarized control loop (so-called "positive" feedback), which does not return to its resting position when an impulse is given but runs to a stop and remains there. If the abovementioned person is not corrected by other influences, he will hopelessly develop into a misanthropist, eccentric, or the like.

The "contagiousness" of feelings grows stronger as a greater number of people are gathered (e.g., due to conformation or fanaticism). The feelings of masses of people are extremely unstable and potentially explosive (e.g., rioting may occur among peace demonstrators); with clever manipulation, the masses can easily be influenced to disregard logical reasoning (e.g., this was seen with the Reichstag vote on total war).

2.5 Will Area

Significance: The area of "wanting" involves energy expenditure, endurance, assertiveness, diligence, etc.

Consciousness: To understand human behavior, a distinction must be made between

- *conscious will* (controlled by the mind) and
- *unconscious will* [comes from deeper layers, e.g., from earlier experiences or innate behavior (instinct)].

The conscious area of the human mind is much smaller than we generally assume. Many of our actions are unconsciously controlled and are subsequently "legitimized" by our consciousness (see hypnosis attempts and the postrationalization of past decisions).

2.6 Experience Area

Significance: This area includes experiences, disappointments, shock, and even dreams.

Engrams: Impressions that reach the experience layer remain permanently effective (often unconsciously) and are therefore called "engrams" (they are firmly inscribed in the mind). Engrams shape behavioral patterns and thus help shape one's personality. For example, disappointment usually generates aggression, even against people who have nothing to do with its cause. Conversely, human beings learn most through experiences of success. These experiences are just as important for one's own further education and personality development as for the act of leading (motivating) others: Recognition is much more useful than blame. However, it also requires constructive criticism as an opposite pole.

2.7   Drive Area

Significance: Drives are located in the deep layer (essence) and are therefore unconscious.

Cardinal drives: Drives can be structured in different ways. The categorization of three cardinal drives is helpful (see Table 1).

Drive development: The three cardinal drives are equally latent. The strongest drive is always the one that is acutely the least satisfied. In the case of equal satisfaction, the oldest drive in terms of developmental history is the strongest: contact drive, then prestige drive, and then possessive drive.

Table 1   The three cardinal drives. Based on Schleip (see Ammelburg, 1993)

Cardinal drive	Examples	Excessive forms	If neglected
Possessive drive (wanting)	Hunger, thirst for knowledge, pursuit of profit, striving for a better standard of living	Greed, avarice, envy, gluttony	Dissatisfaction, social tensions
Prestige drive (being)	Striving for recognition and social prestige, ambition, educational aspirations, need for beauty	Egocentricity, imperiousness, lust for power, bragging	Dissatisfaction, feeling of inferiority (and thereby exaggerated claims)
Contact drive (striving for community)	Sex drive, play instinct, need for sociability, need to lean on others	Pleasure-seeking, herd instinct, massification	Dissatisfaction, uptightness (social isolation)

Drives and motivation: To the extent that a human succeeds in satisfying his cardinal drives, he has the feeling of success (sense of achievement). The slow increase in feelings of success creates motivation. Cardinal drives are thus the source of possible feelings of success, motivation, and, thus, leadership. If a drive remains unsatisfied, this cannot be compensated for by a stronger satisfaction of the other drives (money does not replace, e.g., missing recognition). Nevertheless, the mechanism of drive satisfaction is critical. The psychological "zero level" gradually levels off to the state of the average satisfaction of the three cardinal drives. From there, the satisfaction of the drives and the personal desires are "measured." Prosperity leads to increasing demands and time of need to adaptation (fortunately). Therefore, the following is considered a general rule of life:

4 Drive satisfaction

Every human being seeks to satisfy his drives. However, in the long run, it does not make him happy but only creates new demands.

All those who seek their happiness in the fulfillment of wishes or preach on this path succumb to the pull of this "screw without an end." Someone can become "happy" only by voluntarily, consciously, and, thus, genuinely renouncing things he could or would like to have, that is, by becoming aware of the processes and changing his own "zero level" (this can be difficult). Ghandi said, "One becomes rich only through things one does not desire."

3 Types of People

3.1 Basic Types

Body systems: When an egg cell of a higher living being develops, three cell clusters are initially formed. From this, the three main body systems, which make such a being viable, are developed:

- *Motor system:* This includes muscles, bones, joints, the spinal cord, and the motor nervous system. It enables movement, attack, and escape (it is *active*, i.e., an acting system; mainly an *energy system*).
- *Vegetative system:* This includes the digestive system, connective tissue, and the vegetative nervous system. It absorbs food and turns it into energy for the body (it is *passive*, i.e., an absorbing system; mainly a *substance system*).
- *Central system:* This includes the brain, senses, and central nervous system. It receives information and processes it; it controls and coordinates bodily functions (it is a *superordinate* system; an *information system*).

Physique types (constitution types): The three body systems described above form and shape the body of a human being, which can be described using physique types:

5 Physique type

The appearance of a person is shaped by the system that is most highly developed. Basically, everyone is a mixture of these systems.

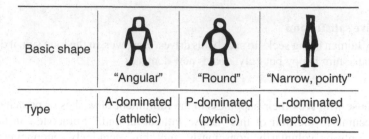

Basic shape	![Angular]	![Round]	![Narrow, pointy]
	"Angular"	"Round"	"Narrow, pointy"
Type	A-dominated (athletic)	P-dominated (pyknic)	L-dominated (leptosome)

Fig. 3 Physique types in extreme principle representation. Based on Kretschmer (1977)

Typical (extreme) manifestations according to Kretschmer (1977) are shown in Fig. 3:

- *Motor* system is dominant: A type (*"athletic"* type).
- *Vegetative* system is dominant: P type (*"pyknic"* type).
- *Central* system is dominant: L type (*"leptosome"* type).

In addition to Kretschmer (1977), a number of other researchers, for example, Sheldon (1963), have defined and investigated similar types.

Type classification: The dominant principle of a type can often be determined from a person's external properties. However, there are restrictions to be considered:

- Since every human being is a mix of types, individual characteristics may well deviate from the dominant principles of a given type. These characteristics usually point to the second strongest principle.
- Illness can distort one's external appearance (e.g., emaciation can simulate an L type body, somebody with a metabolic disorder and bloating can simulate a P type body).
- Mixed types are sometimes difficult to determine, especially in mixtures of a dominant L type with an A or P type as the second strongest component. In some people, all three systems are equally balanced.

Following this, Sheldon (1963) classifies people using a diagram with three cornerstones (see Fig. 4). Serial studies have revealed different common physiques for women and men.

3.2 Physique and Character Traits

Connection between type and characteristics: Centuries of experience and numerous research studies show that there are clear connections between the physique type and other characteristics, for example, the way one moves, reacts, and thinks

Fig. 4 Classification of human types in a diagram with the three basic types: A, P, and L. Based on Sheldon (1963)

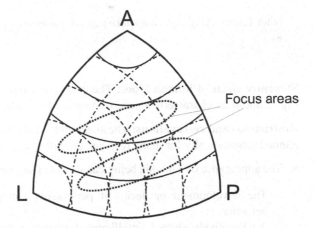

(even if these connections are partly denied in modern psychology). The following consideration makes this fact understandable:

> **6 Reaction type**
>
> In a stressful situation, a person will instinctively rely on his internal system that is most highly developed.

For example, in the event of an impending catastrophe, the following is to be expected:

- An A type (acting system) person will immediately run off and take action.
- A P type (receiving system) person will gather other people together, collect opinions, and discuss possible measures that can be taken.
- An L type (information-processing system) person will withdraw and work out a disaster plan.

In general, it is impossible to say which action is the "right" one. Depending on the situation, anyone can be "wrong" some times and "right" at other times; that is, the system is basically *value-free*.

Consistency of the essential characteristics: The law of the continuity of characteristics (see Fig. 1) is the generalization of the connection between mixed types and characteristics (see Ammelburg, 1993).

> **7 Continuity of characteristics**
>
> If it is possible to determine a person's "type" in one of the layers of the "onion model" in Fig. 1, then conclusions can be drawn about characteristics in the

other layers. All these characteristics are value-free (see Guiding Rules 6 and 8).

Structure circle of human types: The three basic types, A, P, and L, are shown in Fig. 2, and their characteristics are summarized in Table 2.

Restrictions and reservations: The statements of the theory of types must never be applied schematically for many reasons, including the following:

- The appearance of a human being can deviate from these preformed dispositions.

  - The development or decay of personality can greatly change someone's behavior.
  - An intuitively shaped, intelligent thinker can gain a significant amount of logical thinking skills.

- In all cases where collective laws apply, it is not possible to draw conclusions about the behavior of the individual from the collective statement with certainty.
- A typology can basically provide only limited statements, not only because it represents reality only in a model-like or highly simplified way but also because it represents only a limited section of reality. It is as if someone imagines a whole (the human being) as a sphere and makes a cut through the sphere to find a structure. The statements, especially the "poles," are valid only in this sectional plane. Other theories of "types" lay down different "sectional planes" and find different, though related, results:

  - the three types of psychophysiology according to Huter (see Kupfer, 1989) and
  - the nine types of enneagrams, also based on a threefold division.

Self-knowledge and polarity: To exclude as much misuse and misinterpretation as possible, the following should be expressly pointed out:

8 Evaluation of characteristics

All statements about human qualities derived from the typology are completely *value-free*. Each characteristic can have both "positive" and "negative" expressions.

This is the same as the color of one's eyes or the height of one's body: these facts do not contain any valuation (see Fig. 2).

Table 2 Characteristics of the three physique types: A, P, and L. Based on Schleip (see Ammelburg, 1993)

Characteristic		Dominating principle		
		A (athletic)	P (pyknic)	L (leptosome)
System	Focus on	Energy	Substance	Information
	Character	Active, operating	Passive, receiving	Superordinate, controlling
Body	Organs	Muscles, bones, joints	Digestive system, connective tissue	Brain, senses
	Nervous system	Motoric, spinal cord	Vegetative	Central nervous system
	Growth	Angular, muscular, broad	Round, soft, "padded"	Slim, delicate, lean
	Emphasis	Chest	Belly	Head
Head	Form	Broad skull, angular	Round skull	Long skull
	Emphasis	Center (pelvis)	Bottom (double chin)	Top ("thinker's forehead")
	Hair	Thick, frizzly	Soft, wavy	Thin, stringy
	Baldness	Rarely	Round bald head, rear	Bald sides, front
	Nose	Strong, prominent	Fleshy, blunt	Narrow, pointed
	Chin	Angular	Round	Pointed
	Voice	Loud, strong	Soft, melodic	Restrained, soundless
Hands	Structure	Strong, sinewy	Fleshy, soft	Narrow, delicate
	Finger	Broad, angular	Short, thick	Long, pointed
Presentation	Movements	Taut, angular	Calm, soft	Hesitant, nervous
	Writing	Angular, strong	Round, fluent	Thin, steep
	Expression	Simple, succinct	Descriptive, wordy	Careful, precise
Mental structure	Emphasis on	Will	Feeling	Mind
	Joy in	Strength	Pleasure	Problem-solving
	Willingness to	Attack	Compromise	Opposition
	External contact	Extroverted in action, introverted in feeling	Extroverted	Introverted
Behavior	Disposition	Energetic	Compensating	Critical
	Tendency to	Enforce	Adapt	Separate
	Propensity for	Practical things	Personal things	Formal things

(continued)

Table 2 (continued)

Characteristic		Dominating principle		
		A (athletic)	P (pyknic)	L (leptosome)
	Appearance	Fearless	Kind	Reserved
	In case of trouble	Wants to immediately do something	Wants to commune	Wants to retreat and reflect
Leadership	Manner	Energetic	Balancing	Organizational
	Area of responsibility	Pioneering work, implementation	Group work, team, contact	Individual work, research and development

9 Self-knowledge

The value of the theory of types lies, above all, in recognizing one's *own* essential structure and thus preventing one's own way of thinking and acting from being unconsciously regarded as the "right" way and devaluing those toward the opposite pole.

Self-knowledge and becoming conscious are the essential goals of human life; this is how humans differ from animals. The inscription "know thyself" was written above the Greek temple of the oracle at Delphi. The opposite pole is what is more or less missing from one's own personality and thus should be taken in (learned). From this follows an important rule of life:

10 Opposite pole

Everything that someone *rejects* or is afraid of belongs to the opposite pole and should be relentlessly presented to him repeatedly until he has learned to understand and accept it.

Pole and opposite pole ultimately form a unit, and the tension between them leads to development (the way a battery provides energy). It causes the attraction of the poles (after all, opposites attract). Fascination (attraction) and horror (rejection) are only gradually different manifestations of the same effect. Among other things, the truth of this can be seen in the effect of famous comic pairs, for example, Stan & Ollie and Max & Moritz, facing each other in the structural circle.

4 Influencing Factors for Leadership

Significance: The psychological basics from Sects. 2 and 3 are summarized here to provide systematic help for those in leadership positions. A prerequisite for meaningful leadership is the proper handling of the following variables (according to Schleip; see Ammelburg, 1993):

- positive attitude,
- relatively appropriate degree of influence, and
- frequency of influence.

In addition, the five essence areas and the three cardinal drives should be taken into account.

Positive attitude: Without a positive attitude, any attempt at meaningful leadership will be futile. Leadership is possible only under the following conditions:

- *Trust in the person* (decisive); the most important basis for this is fairness.
- *Interest in the matter* (motivation).

Relatively appropriate degree of influence: This can be achieved only through *knowledge of human nature* and the *observation* of the involved parties (types of people) in the sense of a "control loop" (experiences). Influencing too much is just as harmful as too little.

Frequency of influence: Saying something only once does not guarantee that it will stick. For something to remain in one's consciousness, a threefold influence is usually necessary (e.g., lectures, exercise, and tests). Purely intellectual intake and retention are not enough; one must process the experience oneself, and this requires patience: "Patience is the measure of adulthood."

Essence areas: As many of the five inner essence areas (senses, thinking, feeling, wanting, and experiencing) as possible should be addressed simultaneously. For example, the eyes play the most important function of all the senses. This highlights the importance of knowledge of the essence areas and the human types, which in turn shows the value of having a relatively appropriate degree and frequency of influence.

Cardinal drives: Only when the three cardinal drives (possessiveness, prestige, contact) are satisfied in a fairly even way can a human being's own driving forces be coordinated with leadership goals (motivation). Otherwise, dissatisfaction and resistance will arise.

References

Ammelburg, G. (1993). *Die Unternehmens-Zukunft* (4th ed.). Freiburg i. Br.: Haufe.
Endres, H. (1954). *Ich lerne Umgang mit Menschen.* Stuttgart: Fackelverlag.
Kretschmer, E. (1977). *Körperbau und Charakter* (26th ed.). Berlin: Springer.
Kupfer, A. (1989). *Grundlagen der Menschenkenntnis* (29th ed.). Arlesheim: Carl-Huter.
Sheldon, W. H. (1963). *The Varieties of Human Physique.* New York: Hafner.

Successful Teamwork in a Product Development Project

1 Fundamentals

Reasons for teamwork: Teamwork is now an essential element of the internal work structure of any company. The breadth and depth of knowledge has been increasing and continues to grow to such an extent that no single person is aware of all the information in a given field. For this reason, different specialists usually need to work together to solve problems (see Fig. 1).

Requirements for a team: In the so-called *face-to-face group* ("everyone sees the other"), the following requirements apply:

- manageability [5 to 7 (maximum 10) people],
- spatial proximity among the persons involved,
- possibilities for direct communication (without technical aid).

The above group is *not yet a team*. For a team to be formed, the following must be done:

- common goals among the members must be established (acceptance of the goal, will to achieve the goal);
- the norms and rules of the team must be affirmed (see below);
- the members must be integrated into the task building and take over subtasks;
- there must be positive interactions between the members (looks, words, notes, etc.), be they conscious or unconscious actions; and
- the team's success must be valued more than personal success.

Types of teams: Some essential types of team formation are as follows:

- *Full-time team:* This type of team is exclusively occupied with one (usually extensive but limited) task; the members are released from all other tasks as far as operational conditions allow. Permanent teams are not practical (due to the risk of "worn-in tracks," etc.). Therefore, the members should change regularly, as

J. Schlattmann and A. Seibel, *Structure and Organization of Product Development Projects*, https://doi.org/10.1007/978-3-030-81046-7_9

Fig. 1 Synergetic effect through teamwork. Based on Krehl and Ried (1973)

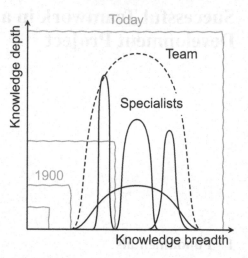

a team with appropriate flexibility creates the necessary conditions for creative product development work.

- *Part-time team:* In this type of team, the members continue to carry out their normal activities in addition to their work as a team. The essential factor here is the frequency of their cooperation. If they work together happens too infrequently, memory and positive attitudes are lost, and the task fizzles out or is neglected. If teamwork is scheduled too often, the actual work may be neglected, and the motivation for teamwork as a whole may be lost (due to fatigue). The decisive factor here is finding the right balance of teamwork. The experience of the team leader is of great importance in this regard.

Role of the team leader: More detailed analyses show that the fulfilment of the criteria for successful teambuilding is largely determined by the team leader. His competence and motivation can determine the success or failure of teamwork. Special knowledge of human nature with leadership qualities as well as the technical and social competence of the team leader are important prerequisites. Ideally, these prerequisites should also be found among the team members because it is well known that the strength of a team—like a chain—is determined by the weakest link.

Organizational conditions: Team success is closely linked to the following organizational conditions and requirements:

- *Task:* This must be clear and precise and should be set by, for example, corporate management before actual work begins.
- *Goal statement:* This must be checked by the team and corrected or further specified if necessary.
- *Management:* Sometimes, it is appropriate for someone other than the project leader to manage the team. The project leader wants, above all, to develop "his product" and may be afraid to address conflicts within the team (see below).

- *Estimation of economic efficiency:* For planned teams, estimates of the effort required for meetings, work outside of meetings, and, if necessary, implementation must be carried out. The amortization period is usually 5 to 6 (or up to 12) months; longer periods should be selected with more caution.
- *Scheduling and tracking:* This is absolutely necessary and can be ensured and simplified by networked information systems if necessary. This also applies to the following point with regard to the inclusion of corresponding databases or similar.
- *Information:* The team must receive all necessary information in a timely manner.

Advantages of teamwork: The advantages of teamwork appear mainly in the following areas:

- *Synergistic effect:* A good team is stronger than the sum of its individual members' skills (see Fig. 1). The team members support and complement each other (in terms of their knowledge, types of thinking, etc.). Often, individuals' skills have a multiplicative rather than an additive effect (then, as is well known, no factor must be too small), which may therefore influence the selection of the group and its team members.
- *Communication:* Thoughts must be verbalized and expressed. This often leads to greater clarity in the product development process and to the replacement of unconscious prefixations, for example, by leaving "worn-in tracks."
- *Motivation:* Good teamwork has a motivating effect because the individual uses the opportunity to excel or because he feels committed to the team (out of a desire for self-affirmation).

Disadvantages of teamwork: Successful teamwork also requires having knowledge about the possible disadvantages of teamwork:

- *Personal friction* can lead to the paralysis of the group.
- Having *too many members* can make the group cumbersome and can negatively affect decisions because "brilliant" ideas are suppressed just as much as "poor" ones; then, only mediocre results are achieved.
- *Dominance by one group member* can lead to the suppression of those with other opinions (especially creative "lateral thinkers"); thus, the synergetic effect is lost.
- *Too-good team climate:* One-sided group thinking can also cause individuals to hold back critical ideas to avoid endangering an overly positive climate. This can lead to the group overestimating its competence and making risky decisions.

2 Behavior of Team Members

Group-dynamic effects: In addition to the human conditions among the team, group-dynamic aspects play an important role. They can lead to, among others, the following *positive effects*:

- *Mutual help* with commitment to the team: Team members feel committed to the team goals, which sometimes go beyond their personal goals. Contacts are made and friendships develop among team members.
- *Increased trust* through general predictability and appropriate behavior among the team members: Despite their individual differences, the team members treat each other as partners and try to use the strengths of others to compensate for their own weaknesses. Communication and cooperation take place at a high level.
- *Personal development:* Team members with increasing popularity can grow into leadership roles; at the same time, they gain high goal awareness through knowledge of their own role in the team. This also increases the performance-oriented use of methods as tools.
- *Effective communication:* This helps team members obtain necessary information and support the work of others.
- A fertile breeding ground for *creative action and development* is built.
- *Improving the social and communicative competence* of team members: Team members develop a deeper commitment to the company and to themselves.

However, the *negative effects* should not be underestimated either:

- *Development of tensions:* Disputes over trivialities may occur, and there may be a shift in solidarity to various parts of the team, for example, through clique formation.
- *Growing insecurity:* This may lead to frustration among individual members.
- There may be increasing *disinterest*, a lack of commitment, etc.

These negative influencing factors disturb the team, possibly even causing its incapacity to work. Thus, herein lies the essential task of the team leader.

Teambuilding process: Teambuilding usually takes place in several steps:

- *Orientation step:* At the beginning, the team is not yet a real team; the team members try to find themselves, frequently with a wait-and-see attitude (because the team still lacks a foundation of trust) and careful mutual sensing. First members of the team give impulses and seek team leadership. The purpose and the mission of the team are clear but still not very motivating because many people are not used to typical team procedures and methods.
- *Determination step:* The first feelings of trust develop among team members, whereby common goals become clear; often, however, confirmation and guidance from outside the team are still necessary. Slowly, a clear development of the team roles takes place through attempts at self-assertion among the team members and efforts to "control" the team. Here, the first clear signs of synergetic effects appear.
- *Teamwork step:* A strong common will to accomplish tasks together as well as high levels of mutual trust and personal investment in achieving the team goal develop (trust is more stable than mistrust). Role-finding in the team is usually completed, and the team mechanisms and methods used are generally fluid. Communication is versatile, and there is a high willingness to cooperate with other teams and

external actors, as well as a growing identification with and commitment to the company and oneself. Restrained team members largely become more involved.

Role allocation among the team: After a start-up period of a few sessions, the following roles are often defined (though not necessarily):

- *Informal leader:* This is the most widely accepted member; the selection arises unconsciously and without conscious manipulation. If the informal leader places himself outside the values and norms of the team (e.g., through claims to power or vanity), he forfeits his role. This then leads to difficulties in the team. The informal leader does not necessarily have to be the team leader (even if this would be convenient); the latter may also have to take a stand against group norms, which would inhibit him from being the informal leader. An example of this would be a works council that has to represent negotiated compromises.
- *Prudent advisor:* This person is accepted and appreciated mainly because of his professional qualifications and the radiated calmness with which he also masters difficult situations. He is a good team leader.
- *Scapegoat:* This is the weakest team member and the one who is blamed for all the mistakes and failures of the team. This person requires special protection from the team leader.
- *Optimist:* This person helps develop a positive atmosphere with a good temperament and sense of humor. He is vital for the team, especially in case of difficulties, unexpected problems, etc. He can resolve blocked situations and thus initiate new ideas.
- *Pessimist:* This person is also very important because he can foresee mistakes, difficulties, etc.; however, he must not be too dampening (particularly in the green phase).
- *Normally active employee:* This person performs his work calmly and reliably without being particularly prominent. He is valuable for the team, especially for working on tasks between meetings (e.g., implementation, elaboration).
- *Passive employee:* This person does not cooperate with others and is obviously being misused. The team leader must clarify whether the cause is personal or professional problems or a lack of team suitability (this is not a disqualification).

The ideal team member: In addition to the coverage of various required fields of knowledge and professional qualifications, the essential characteristics of the team members are decisive. Ideal people do not exist; therefore, the following list of essential characteristics is a desirable goal. However, a single poorly selected member of the team can possibly derail the efforts of the whole team. The "ideal" team member is:

- adaptable (not subordinate),
- willing to accept other people and ideas,
- able to think abstractly and analytically (without forgetting practice),
- unprejudiced (nonjudgmental and not ignorant),
- familiar with team rules and willing to accept and apply them,

- assertive (not stubborn),
- humorous (not silly),
- nonharming and nonvulnerable,
- balancing (does not need to be balanced himself),
- positively charismatic in engaging with others,
- collegial and natural in behavior (no distance from superiors), and
- willing to fully support compromises.

Team values and norms: By accepting the team goal, common interests and values ("we" instead of "I") are created—consciously or unconsciously. This leads to norms of behavior that the team imposes on the individual, for example,

- punctual and comprehensive completion of assigned tasks;
- punctual attendance at team meetings, no absenteeism, no assigning third parties to stand in for someone, no leaving the team, etc.;
- avoiding negative, emotionally colored statements about team members, other employees, or the company and its products;
- finding a consensus on a common and equal allocation of shares in inventions among the team members (an exception being the consensus of all team members on another allocation scheme);
- no personal criticism of team members

  - by commissioning third parties;
  - in the presence of third parties;
  - via telephone, e-mail, or other correspondence; but

only in personal conversation.

Criticism can be accepted only if it is presented

- in private;
- eye to eye;
- in a serious, thoughtful, and calm manner;
- with one's own conviction; and
- with a positive basic attitude (not a destructive one).

3 Team Meetings

3.1 General Information

Significance: If team coordination or work is carried out in the context of meetings, the necessary conditions for this must be taken into account. Examples of this are:

- *the leadership ability of the leader* (decisive), and

- the ability of the *participants* to discuss *factually* and with *psychological* empathy (this includes appropriate knowledge).

Types of meetings: A distinction is made according to the purpose of the meeting (see Ammelburg, 1976, 1993):

- a *problem meeting* is used to solve problems; that is, the result is open (here, this is the most important type); and
- a *goal meeting* has a given goal; the purpose is to determine the path to the solution; possibilities are

  - an *informational meeting* to impart knowledge (change in consciousness),
  - a *motivation meeting* to influence behavior, and
  - a *teaching discussion* (combination of the two).

According to the type of *organization*, the following categories of meetings can also be distinguished:

- a *scheduled meeting* (on a regular basis or, if required, with a written invitation) and
- an *ad-hoc meeting* (spontaneous or due to a sudden event, e.g., damage event).

Meeting room: The meeting room is important for the atmosphere of the meeting. Requirements for the meeting room include the following:

- A quiet, closed, *friendly* design.
- *No telephone* or mobile phone; if there is a phone in the room, unhook the receiver or switch off the phone. In an emergency, a messenger can come.
- An *appropriate size* relative to the number of participants (too small means bad air and thus fatigue, too large causes a feeling of being lost).
- A *table* where everyone can sit *"equally"* and see each other. A circular table is ideal, and octagonal, hexagonal, and square tables are the next best. The more "table poles" are formed, the more tensions are likely to arise (see Ammelburg, 1976, 1993). For example, in the case of long, U-shaped tables, "executive committees" or "opposition" or "contra" parties often sit opposite each other at the greatest possible distance.

3.2 Organization and Procedure

Preparing for a meeting: The following steps are primarily the responsibility of the meeting leader:

1) *Define the participants* if they are not already determined. A reasonable number of participants is approximately 5 to 12; if there are more than 20 participants, a real meeting will usually not be possible.

2)   *Select the date* if it is not fixed by rotation. This should be delegated (e.g., to an assistant); if necessary, Doodle or a similar schedule system can be used.
3)   *Invite participants* in writing at least five working days before the meeting. The invitation should contain information on the following:

- *What:* The name of the meeting and an agenda; if necessary, explanatory keywords or information can be given, but no anticipation of the meeting result should be shared.
- *Who:* The invited participants.
- *How:* Documents relevant to the agenda items, if necessary.
- *Where*: The building and room.
- *When:* The date and time.

4)   *Create a time schedule.* Estimated times for the individual agenda items, especially for longer meetings, should be specified on a flipchart or using a visual tool; this has an accelerating effect if the allotted time is exceeded and is necessary when inviting guests to address individual agenda items.
5)   *Prepare content.* The meeting leader should briefly discuss important points with the main speaker (e.g., the applicant) in order to be informed about problems and subitems (it is important not to indicate the anticipated result of the meeting). Recurring meetings should be neither too frequent (otherwise, fatigue will result) nor too infrequent (otherwise, memory and attitude will decline).

Sequence of a meeting:

1)   *Welcome:* This should be a friendly and polite greeting by the meeting leader (or the person who sent the invite); otherwise, the "positive" attitude may be lost from the beginning.
2)   *Opening formalities* include the following:

- determining whether there is a *quorum* (if necessary);
- selecting a *meeting minute taker* (if not already determined); and
- making additions to the *agenda* (by the leader and the participants—these should be specifically asked for); add items to the agenda and obtain the consent of the participants as necessary (holding a vote may be required), but *under no circumstances should additions be discussed.*

3)   *Address the first agenda item:*

- *Each problem should be presented* by the person who brought it up (this should be short, concise, and supported by illustrative material, drawings, etc.).
- *Name the goal* (but do not anticipate the result). This should be done by the speaker or leader; if necessary, ask specific questions to open the discussion.
- *Discussion* (see Sect. 3.3)*:*

  – Consider requests to speak (via raised hands).

- First listen and understand and only then, if necessary, express a different opinion.
- Allow participants to finish, but limit lengthy speeches (see below).
- If necessary, repeat the core question so as not to lose sight of the goal.
- Either ignore additional problem points that arise (ask for agreement) or write them down and put them aside for later consideration.

- Filter out the *result* from the opinions (as far as possible); formulate it, and note it down. Obtain approval or suggestions for changes; vote if necessary.
- Determine *decisions* precisely and unambiguously:

  - *What* is to be done?
  - *Who* will do what?
  - *How* will it be done (procedure, tools and methods, budget, etc.)?
  - *Where* (for what) will it be done?
  - *When* does it have to be finished (deadline)?

4) Treatment of the next agenda items as in Point 3.
5) *Breaks:* Asking participants to attend a meeting lasting several hours without breaks is not only an unreasonable demand but also pointless. As a result of fatigue, concentration and positive attitudes rapidly diminish. Therefore, the following advice is offered:

---

1 Meeting breaks

After two hours *at the most*, a break of about *five to ten minutes* should be taken, even if a discussion is still ongoing.

The meeting will not take longer (especially if the break times are strictly adhered to). If there are no breaks, the participants will leave the discussion at random and will be absent for a while. During the break, tensions subside (there is a bit of a reprieve); often, new suggestions and approaches to solutions arise afterwards.

6) *Conclusion: Positive* closing words from the leader (at least appreciation is expressed for the participants' commitment), even if the goal has not been achieved. Otherwise, a negative attitude remains and burdens the next meeting right from the start.

If any agenda items could not be discussed due to lack of time, they are usually placed at the beginning of the next meeting.

Special features of the ad hoc meeting: An ad hoc meeting is called at short notice for sudden reasons (e.g., a damage case, production problem, customer complaint). In these cases, the following should be noted:

- *Preparation:* The person bringing the problem up should clarify the date and place with the affected participants by telephone and note the *main topic.*

- *Greet and lead:* A friendly greeting should be given (as above) by the inviting person. The essential action here is selecting someone to lead. If nothing can be agreed upon, the inviting person takes over.
- *Minutes-taker:* Equally importantly, someone has to take notes (keywords are usually sufficient; see below).
- *Agenda:*

  - *The main topic* and, if necessary, further topics are named by the leader.
  - *Further topics* can be brought up by participants (as above; only mention them, but do not discuss them yet).
  - *Note* agenda items in writing; if necessary, use a flipchart, overhead, or whiteboard (leader and meeting minute taker).
  - Ask about *priorities* and determine the order of the agenda items; if necessary, define time limits (however, avoid major discussion).
  - Obtain *consent* to start the meeting.

- *Conduct the meeting* as described above:

  - address the individual agenda items,
  - take breaks in longer meetings, and
  - reach a positive conclusion.

Minutes: Taking minutes is absolutely necessary; otherwise, the meeting will have been useless. Meeting minutes ensure that the decisions made are carried out, but above all, that it is later possible to understand what has been decided and why. These can be very short (keywords are enough).

2 Minutes

The minutes must contain at least the following information:
- *basic data* (name of the meeting, day, time, participants, agenda),
- *purpose* (keywords),
- *result* (keyword summary),
- *decisions* (what, who, how, where, and when), and
- *the name of the minute taker.*

The minutes do not include details on the course of the discussions. Film recordings of the meeting should not be made.

3 Writing up the minutes

The minutes should be written *immediately after the meeting* and reviewed and approved by the meeting leader.

If the minutes are written up later, they will take at least twice as long to write. After a few days, the meeting minute taker will no longer have the essential facts in his head.

4 Distribution of the minutes

The minutes should be sent to the participants *immediately*. Otherwise, the action items will not be executed on time, because everyone relies on the minutes to track what they have agreed to take responsibility for.

3.3 Behavior of the Meeting Leader

Significance: The meeting leader has the most important and most difficult task in the meeting. Experience and, if possible, training in knowledge of human nature and meeting techniques are important. However, for the other participants, this knowledge is also helpful.

Guiding rules for meeting techniques: The following guiding rules summarize some essential aspects of meetings using keywords. They may serve as an introduction to further study the topic.

5 Equivalence

The leader should regard and address the participants as equal partners.

This includes being tactful, polite, friendly, decisive, and consistent. Politeness oils the gears. Genuine friendliness is even more helpful; it corresponds to a general attitude or way of being and indicates that someone is sincere with himself and others. False friendliness, on the other hand, is a scam that is easily seen through. Determination and consistency (not hardness and stubbornness) are essential, but they can be realized in a polite and friendly way.

6 Moderator

The meeting leader is a moderator, that is, someone who gently directs but does not control or manipulate. Although he is often asked questions, they are generally directed at the entire group and should therefore be passed on to all participants.

This means that the leader talks as little as possible and holds back his own opinion at first because the problem should be solved by the group. Before moving on, the leader should check with the person asking the question to see if he considers the question sufficiently answered.

7 Order of speech

Regulate the order of speaking (e.g., using hand signals), even in small groups.

This also means ensuring that all attendees participate; if necessary, follow up discussions by asking questions.

8 Questions

Formulate questions clearly. When asking a question to a specific participant, first state his name and then ask the question. If the question is open to all, be patient and wait.

By allowing space for silence, embarrassment is avoided in the case of mental "absence" (this naturally happens all the time because our mind is constantly moving from one thought to another). In the latter case, no one feels directly addressed; everyone can think about it until someone finally takes the floor.

9 Formulations

Clumsy formulations where it is uncertain whether the others have correctly understood the speaker should not be corrected but tactfully expressed in as a summary of the speaker's words: "Did I understand you correctly that…?".

In this way—as long as it does not happen too often –, the offense of the other person can be avoided (some thinkers occasionally find it hard to use words to communicate effectively).

10 "Wrong" views

If the leader thinks that a view is wrong, he should not object to it immediately but should let the group comment on it.

Only if there is no reaction can the leader express his own opinion.

11 Interactions

Free interactions between participants, that is, without requests to speak, can be allowed to continue as long as they remain on the topic and do not get out of hand and there are no urgent requests to speak. The result of such interactions should be summarized briefly if necessary.

12 Digressions

If the discussion deviates from the goal, the leader must intervene tactfully; if necessary, any new topics that arise can be noted as a further item on the agenda.

13 Long speeches

Long-winded speeches should be interrupted with caution and skill at a convenient point.

This can be difficult to do, as regardless of how it is done, the interrupted person may be angry. However, if such speeches are allowed to continue, other participants may become angry. For example, the general correctness can be emphasized, but at the same time, reference can be made to other requests to speak or the progressive time.

Dealing with meeting participants: Meeting participants may be perceived and dealt with by the meeting leader in the following ways (see Ammelburg, 1993):

- *The quarrelsome person:* Remain objective and calm and do not get involved in arguments. This person cannot be refuted by discussion, so the "blind spot" tactic may be used.
- *The positive person:* This person provides support for the meeting. Let him summarize the results and consciously engage in the discussion.
- *The know-it-all:* Let the group comment on his assertions.
- *The talkative person:* A loquacious person. Tactfully interrupt and limit his speaking time.
- *The shy person:* Ask him easy questions and raise his self-confidence through praise and recognition.
- *The rejector:* Awaken his ambition and acknowledge and utilize his knowledge and experience.

- *The aloof person:* This person is uninterested in the meeting. Ask about his work and solicit an example from his field of interest.
- *The sublime person:* This person is a big shot. Do not criticize him; use the "Yes, but" technique.
- *The questioner:* The sly fox. He may try to trick the meeting leader. Pass his questions to the group for comment.

Parliamentary rules: For a meaningful and fair discussion, the meeting participants, especially the leader, must know and observe the most important parliamentary rules:

- *Voting mode:* Without exception, a vote must ask the following:

  1) Who is *in favor*?
  2) Who is *against*?
  3) Who *abstains*?

Omitting Step 1 is inadmissible because inattentive people are counted as being "in favor."

- *Request for a secret ballot:* This request must always be granted, even if only one person makes it (without voting).
- *Distribution of words:* This is done by the leader in the order of the requests (following the speaker list).
- *Request to speak "directly to it":* This occurs immediately after the current speaker. This may contain only a short addition (no extensive new contributions) and should occur only *rarely*.
- *Request to speak "on a personal note":* This should occur only after the factual discussion of an item has been concluded. This is only done to correct personal assertions, incorrect information, etc.
- *Request to speak "on a point of order":* This is considered *immediately* after the current speaker finishes. It serves *exclusively* to make one of the following three requests.
- *Request to "close the list of speakers":*

  1) The leader names the final speakers.
  2) *One* speaks *in favor* of the request (generally the proposer).
  3) *One* speaks *against* the request (if desired), but there is no further discussion.
  4) *Voting* (see above) occurs. If the proposal is accepted, only the named individuals will speak.

- *Request to "close the debate":*

  1) *One* speaks *in favor* (as before).
  2) *One* speaks *against* (if desired).
  3) *Voting* occurs. If the proposal is accepted, no further discussion takes place (in particular, no "funeral orations"); if necessary, the item discussed can be voted upon.

- *Request for a "transition to the agenda":*

  1) *One* speaks *in favor*.
  2) *One* speaks *against*, if necessary.
  3) *Voting* occurs. If the proposal is accepted, the item is removed, and the meeting transitions to the next agenda item.

References

Ammelburg, G. (1976). *Konferenztechnik*. Dusseldorf: VDI.
Ammelburg, G. (1993). *Die Unternehmens-Zukunft* (4th ed.). Freiburg i. Br.: Haufe.
Krehl, H., & Ried, A. P. (1973). *Teamarbeit und Gruppendynamik*. Karlsruhe: Krehl + Ried, Neue Management Methoden Verlag.

Index

© The Editor(s) (if applicable) and The Author(s), under exclusive license
to Springer Nature Switzerland AG 2021
J. Schlattmann and A. Seibel, *Structure and Organization of Product Development
Projects*, https://doi.org/10.1007/978-3-030-81046-7

153

Printed in the United States
by Baker & Taylor Publisher Services

Printed in the United States
by Baker & Taylor Publisher Services